Periodicity and the s- and p-Block Elements

N. C. Norman

School of Chemistry, The University of Bristol

Series sponsor: **ZENECA**

ZENECA is a major international company active in four main areas of business: Pharmaceuticals, Agrochemicals and Seeds, Specialty Chemicals, and Biological Products.

ZENECA's skill and innovative ideas in organic chemistry and bioscience create products and services which improve the world's health, nutrition, environment, and quality of life.

ZENECA is committed to the support of education in chemistry and chemical engineering.

OXFORD   NEW YORK   TOKYO
OXFORD UNIVERSITY PRESS
1997

Oxford University Press, Great Clarendon Street, Oxford OX2 6DP

Oxford   New York
Athens   Auckland   Bangkok   Bogota   Bombay   Buenos Aires
Calcutta   Cape Town   Dar es Salaam   Delhi   Florence   Hong Kong
Istanbul   Karachi   Kuala Lumpur   Madras   Madrid   Melbourne
Mexico City   Nairobi   Paris   Singapore   Taipei   Tokyo   Toronto
and associated companies in
Berlin   Ibadan

Oxford is a trade mark of Oxford University Press

Published in the United States
by Oxford University Press Inc., New York

A catalogue record for this book is available from the British Library

Library of Congress Cataloging in Publication Data
Data applied for

ISBN   0 19 855961 5

Typeset by the author
Printed in Great Britain by The Bath Press, Bath

Series Editor's Foreword

The s- and p-blocks comprise the most abundant elements and thus are of great relevance as well as fundamental importance. This 'Main Group' chemistry covers a wide variety of materials—from inert gases to the most reactive of alkali metals, including some finer points of the trends between germanium, gallium, and indium in between. These gross and subtle changes often give students problems as they seek to come to grips with these important elements. The theme of this book is to provide a framework for understanding these elements.

Oxford Chemistry Primers are designed to give a concise introduction to all chemistry students by providing the material that would usually be covered in an 8–10 lecture course. As well as giving up-to-date information, this series provides explanations and rationales that form the framework of understanding inorganic chemistry. Nick Norman builds on his earlier volume on the p-block elements and extends his coverage to include the alkali and alkaline earth elements. He uses the intrinsic atomic properties to account for the breadth of the chemistry of these elements.

John Evans
Department of Chemistry, University of Southampton

Preface

This book is a revised version of the Chemistry Primer entitled *Periodicity and the p-Block Elements*. In addition to having updated and substantially rewritten parts of the first book, the most obvious change is that certain aspects of s-block element chemistry are now discussed explicitly and so this group of elements are therefore included in the title. As in the first book, I have sought to provide an overview of some of the important trends (periodicity) found in the properties of the s- and p-block elements and the compounds that they form, and to provide the reader with some simple 'rules of thumb' whereby they might appreciate and better understand these trends. In a book of this length, I have had to be selective and cannot possibly cover all aspects of periodicity (I have not included anything on solubility trends in s-block compounds for example), but I hope that those aspects which I have chosen will be both useful and informative. I make no apologies for the non-mathematical treatment adopted in Chapter 1. If the student wants more, it is available in other texts listed in the bibliography, but I suspect that my qualitative and descriptive approach (in common with a great many other texts) will be perfectly adequate for most of the target audience.

As with the first book, I called upon many people to read the book in draft form and for this I thank my colleague Guy Orpen together with Craig Rice, Edward Robins, Yvonne Lawson, and Siân James who are all current members of my group.

Bristol
February 1997

N.C.N.

Contents

Introduction

A large part of the fascination of inorganic chemistry lies in the almost infinite variety and diversity of the compounds which the elements form amongst themselves. Ionic solids, covalent molecules, metals and alloys, ceramics and polymers are all examples of the types of materials encountered, and nowhere is this diversity more apparent than in the chemistry of the main group elements of the s- and p-blocks, the topic with which this book is concerned. Nevertheless, whilst variety and diversity are fascinating on the one hand, on the other, it may sometimes seem that main group chemistry is little more than a vast collection of disparate facts which can only be dealt with by some monumental feat of memory.

The object of this book is to show that much of the huge body of knowledge which is s- and p-block chemistry can be understood and rationalized on the basis of a few simple and straightforward principles. We start in Chapter 1 with an overview of ideas on atomic structure which lead to an understanding of why the periodic table has the form that it does. The periodic table is the single most important organizing principle in inorganic chemistry and an understanding of its structure is crucial if we are ever to understand the periodic relationships between the properties of the elements and their compounds.

Chapter 2 deals with the periodic aspects of some of the properties of the atoms themselves such as ionisation energies and electronegativity, the latter principle in terms of a possible third dimension to the periodic table. The trends in all of these properties are seen to be closely related and form a foundation upon which we can build an understanding of the nature of the elements themselves, which is the subject of Chapter 3.

Chapter 4 looks at some of the important general features associated with the compounds of the elements such as oxidation states, bond energies and the physical nature of the compounds themselves, such as whether they are ionic, molecular covalent, polymeric or metallic. This provides a basis to explore, in Chapter 5, the specific nature of selected classes of compound, such as the halides, oxides and hydrides, and the factors which determine why these compounds have the characteristics they do. In all cases, the emphasis is on observed periodic trends and the reasons behind why such trends occur.

In Chapter 6 we look at the concept of acids and bases and finally in Chapter 7, at some aspects of structure. A bibliography is provided at the end and, in many ways, this text should be read in conjunction with some of these books since there has not been space enough to provide much in the way of factual information or descriptive chemistry.

1 Atomic structure and the form of the Periodic Table

1.1 Introduction

We should recognise at the outset that the modern form of the periodic table was originally deduced (by Mendeleev and others) from empirical chemical relationships between the elements and that a theoretical understanding came later.

A non-mathematical but more detailed treatment of this subject can be found in Winter (1993). A more empirical, introductory approach, which is also non-mathematical, has been discussed by Gillespie *et al.* *J.Chem.Educ.*, (1996), **73**, 617.

An understanding of the structure or form of the periodic table is essential if we are ever to understand periodicity and for this it is necessary to know something about atomic structure or how electrons behave in atoms. In this chapter we shall examine this topic in some detail and show how these ideas are useful in explaining why the periodic table has the form it does. More detailed treatments of this subject can be found in many of the inorganic chemistry textbooks listed in the bibliography although it will be evident from many of these texts that this topic is often dealt with in quite a mathematical way. Wave-equations and the like are, of course, mathematical by their very nature and a *full* understanding and appreciation of this subject cannot be had in the absence of a mathematical treatment. However, the basic ideas can be presented and understood using a descriptive and pictorial approach and this is the one we shall adopt. This necessarily involves many *ad hoc* assumptions and statements but we will not suffer unduly because of this, perhaps quite the reverse!

Original ideas on atomic structure, which were developed in the early part of this century (after it was deduced that atoms had any structure at all), consisted of a positively charged 'cloud' with embedded, negatively charged, particle-like electrons. This was soon superceded, however, by the Rutherford model in which a central, positively charged nucleus, containing most of the mass of the atom, was surrounded or orbited by the much lighter electrons. The resulting picture of the atom was classical in nature in that the atom was viewed as a miniature solar system in which the nucleus was analogous to the sun and the electrons to the planets, but this simple model, as it stood, was inconsistent with classical physics which predicted that the electrons in such an atom should spiral into the nucleus with concomitant emission of radiation, i.e. atoms should decay!

It was clear, however, from contemporaneous observations in atomic spectroscopy that, for whatever reason, only certain orbits or energies were possible for the electrons, and this formed the basis of the subsequent model of the atom proposed by Niels Bohr (Fig. 1.1). Nevertheless, whilst this model was quite successful in explaining the observed spectroscopic properties of the hydrogen atom, the simplest of all atoms containing only a single electron, it soon ran into problems with heavier atoms containing more than one electron. Furthermore, from a theoretical standpoint, the fact

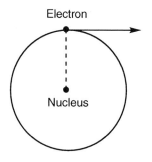

Fig. 1.1 A pictorial representation of the Bohr model of the atom.

that only certain electron orbits or energies were possible had to be introduced on a purely *ad hoc* basis with no clear idea of why it should be so.

An elegant solution to the problem was provided by Schrödinger in 1926 who, building on the earlier work of de Broglie, proposed that electrons in atoms should be considered not as particles but as waves and that they could therefore be described by a suitable wave equation. The important conceptual breakthrough here was that the presence of only certain allowed electron energies (or orbits if we continue to think in a classical sense) is a direct and natural consequence of a wave treatment and no *ad hoc* assumptions are necessary. We can see how this arises in a simple way by considering a one-dimensional model.

A suitable model for a one-dimensional wave-like system is a stringed musical instrument. When the string vibrates, a certain note is heard which has a fixed frequency (or energy; this term will be more appropriate later) and it is therefore apparent that as a result of the string being constrained in some way (the constraints here are the length of the string and its tension), only a certain particular frequency (energy) of vibration is possible.

However, it is important to realise that even without changing the length or the tension of the string it is possible to sound other notes of higher frequency (the so-called higher harmonics), two of which are illustrated in Fig. 1.2, and, as with the frequency of the fundamental vibration, the frequencies of the higher harmonics are also determined by the constraints on the string. Thus, for a string with given constraints, only certain particular frequencies of vibration are possible. At this stage, we can introduce the terms *node* (a point of zero amplitude) and *lobe* (displacement between nodes) and we should then recognise that the higher harmonics have extra nodes symmetrically disposed along the string (in all cases, the two fixed ends of the string are also both nodes); the first harmonic has one extra node, the second harmonic has two and so on (see Fig. 1.2).

These waves which result from the vibrating string may also be thought of as representations of solutions to a one-dimensional *wave equation* for which only certain solutions are possible. In other words we may say that only certain *wavefunctions* (i.e. solutions to the wave equation) are allowed. Furthermore, these various solutions to the wave equation are usually labelled with a number called a *quantum number* which can take only certain allowed values, each one of which corresponds to an allowed solution to the wave equation.

It should now be apparent that this property of only certain allowed frequencies or energies (*quantization* as it is usually called) is a direct and natural consequence of a wave-like system which is constrained in some way, and that for a one-dimensional system there is *one* quantum number, n, which can take various values (i.e. $n = 1, 2, 3$, etc.). Returning now to atoms which are three-dimensional, there will be three quantum numbers (we shall see later that a fourth quantum number associated with the electron itself will also be necessary) and it is these that we will use as labels for the wavefunctions or solutions to the *Schrödinger wave equation* which is used to describe the wave-like properties of electrons in atoms.

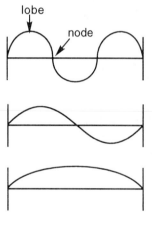

Fig. 1.2 Vibrations of a string fixed at both ends. The fundamental vibration and the first two harmonics are shown. The horizontal lines indicate the position of the string at rest. An example of a node and a lobe are also illustrated.

Before we look at three-dimensional systems in detail, let us review the important points again. The Schrödinger wave equation is used to describe the behaviour of electrons in atoms, i.e. we treat electrons as waves. Only certain solutions to the wave equation are possible, each corresponding to an allowed energy (orbit) for an electron, and this quantization results directly from the system being constrained in some way. The constraints (often refered to as boundary conditions) for an electron in an atom are not as obvious as those for the strings discussed above, but result from the attractive interaction between the positively charged nucleus and the negatively charged electron. Moreover, we obtain not one solution to the Schrödinger equation but a series of many possible solutions each described by a certain set of three quantum numbers. Each one of these solutions or *wavefunctions* (given the symbol ψ) describes a possible state or energy of an electron in an atom and is refered to as an *orbital* (the part *orbit* is leftover from more classical thinking).

Let us now look at the solutions to the Schrödinger equation in more detail and see what they mean. In fact, the solutions or wavefunctions, ψ, refer to the *amplitude* of an electron wave but of more physical significance is the square of this function, ψ^2 which refers to the electron *intensity* (density) or, alternatively, a *probability* of finding an electron in a particular region of space. Thus ψ^2 at any particular point is the probability of finding the electron at that point.

1.2 Wavefunctions for the hydrogen atom

We will start by considering the hydrogen atom as it is the easiest system to look at since there is only one electron and the Schrödinger equation can be solved exactly; it can be used for other atoms with suitable modifications as we shall see later. However, in order to allow us to represent much of what is to follow pictorially, we will want to look at each wavefunction in two parts. An expression which shows how we can partition the wavefunction is shown below, the two parts being the so-called *radial* and *angular* parts.

$$\psi = R_{n,l}\,(r) \, . \, Y_{l,m_l}\,(\theta, \phi)$$

The radial part, $R_{n,l}\,(r)$, depends only on the radial distance, r, between the nucleus and the electron and contains no information on direction or orientation. It depends on the quantum numbers n and l which will be defined shortly. The angular part, $Y_{l,m_l}\,(\theta, \phi)$, depends on the direction or orientation, but not on distance, and is dependent on the quantum numbers l and m_l with the angles θ and ϕ defining an orientation with respect to a suitable coordinate system.

As we have seen, boundary conditions and the spherical, three-dimensional nature of the atom give rise to three quantum numbers and these are given the symbols n, l and m_l. Each of the quantum numbers can take only certain allowed values and a solution exists to the Schrödinger equation for particular allowed sets of these three numbers, any one set of three defining a particular orbital. The names and symbols for these quantum numbers and the values which they can take are given below.

We will not be concerned here with the details of how it is that these particular values for and relationships between the quantum numbers come about, why they are given these particular names or even precisely why it is that three quantum numbers are required. It is the results in which we are interested and the reader is refered to more advanced texts for a fuller discussion of this topic.

(a) n is the *principal* quantum number and is associated with the radial part of the wavefunction. The energy of the orbital also depends on n (see later) but not on l and m_l. The number n can take integral values 1, 2, 3, 4 . . . ∞ but we will be concerned only with the first few.

(b) l is the *subsidiary* or *angular momentum* quantum number and can take values 0, 1, 2, 3 . . . $n-1$, i.e. the possible values of l are dependent on n. The value of l influences both the radial and angular parts of the wavefunction and, in particular, determines the type or shape of the orbital which is usually given a letter designation.

$$l = 0 \quad \text{s orbital}$$
$$l = 1 \quad \text{p orbital}$$
$$l = 2 \quad \text{d orbital}$$
$$l = 3 \quad \text{f orbital}$$

Orbitals are then labelled according to their value of n and the letter associated with l:

$n = 1$	$l = 0$	1s
$n = 2$	$l = 0$	2s
$n = 2$	$l = 1$	2p
$n = 3$	$l = 0$	3s
$n = 3$	$l = 1$	3p
$n = 3$	$l = 2$	3d

Note that orbitals such as 1p and 2d are *not* allowed according to these rules.

(c) m_l is the *magnetic* quantum number and it takes values $-l, -l+1, \ldots$ 0. . ., $l-1, l$, i.e. $2l+1$ values for a given value of l. Thus for $l = 0$, $m_l = 0$ and so there is only one type of s orbital for any given value of n, i.e. one 1s, one 2s etc. For $l = 1$, $m_l = -1, 0, +1$, i.e. three types of p orbital for any given value of n except for $n = 1$, i.e. three 2p, three 3p etc. For $l = 2$, $m_l = -2, -1, 0, +1, +2$, which means five types of d orbital. This quantum number specifies the orientation of the orbital.

The particular numerical values of m_l are not important as far as this discussion is concerned, merely the total number of possible values allowed for any given value of l. Furthermore, we will also not be concerned with the details of the orientation of orbitals with respect to a given coordinate system although we should recognise that this information is very important when considering molecular orbitals.

When we look at these orbitals or wavefunctions in more detail we will see that they contain nodes (just like strings) and there are a few simple rules concerning nodes that are worth remembering. The total number of nodes in any given orbital is equal to $n-1$. Some of these are in the radial part of the wavefunction and some are in the angular part. The number of angular nodes is simply equal to l and the number of radial nodes is therefore equal to $n-l-1$ as we shall see in the following sections.

In the case of the hydrogen atom, the energy of an electron or orbital depends *only* on the value of n and so, for example, the 2s and 2p orbitals have the same energy and are said to be *degenerate*. The energy of the orbitals increases as n increases and this is shown graphically for the first few levels in an energy level diagram for the hydrogen atom in Fig. 1.3. Note that the separation in energy is not constant but that the levels become closer together as n increases. Moreover, the energies are given negative values with $n = \infty$ defined as zero energy.

Now we will look at orbitals in more detail and we will start with the angular part of the wavefunction.

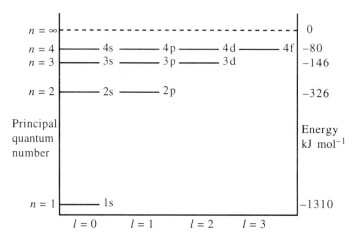

Fig. 1.3 An orbital energy level diagram for the hydrogen atom.

1.2.1 The angular part of the wavefunction

The angular part of the wavefunction reveals how the wavefunction varies as a function of an angle from the origin of some suitably chosen coordinate system and thus determines the shape of the orbital. It is dependent on the quantum number l and, as we have seen above, we can label the types of orbitals according to the value of this quantum number or more usually with the letters s, p, d, f, etc. Remember that the number of nodes in the angular part of the wavefunction (so-called angular nodes) for a given orbital is equal to l. Thus s orbitals (1s, 2s, 3s etc.) where $l = 0$, have no angular nodes and, since there is therefore no angular dependence of the wave function, the orbital is spherical in shape.

For p orbitals, $l = 1$ and there is one angular node. This node is a planar surface and divides the orbital into two lobes of opposite sign. There are three possible orientations for orbitals of this type (three values of m_l) and these lie along the axes x, y and z, and are usually designated as p_x, p_y and p_z. All p orbitals have this shape and there are always three for any given value of n, i.e. three 2p, three 3p etc.

For d orbitals, $l = 2$ and there are now two nodes associated with these orbitals so that each therefore has four lobes. For a given value of n, there are five d orbitals (five values of m_l) which are given the labels d_{xy}, d_{xz}, d_{yz}, d_{x2-y2}, d_{z2}. Pictures of the angular part of the wavefunction for some of these orbitals are shown in Fig. 1.4 but remember that these are two-dimensional representations of three-dimensional orbitals; the s orbital should be thought of as spherical and the p orbitals as two tangential spheres.

As discussed in section 1.1, we will generally be more interested in the square of the wavefunction, ψ^2 and pictures of the square of the angular part of the wavefunction for the s, p_x and p_z orbitals shown in Fig. 1.4, are shown in Fig. 1.5. Note that the s orbital is still represented as a circle (sphere in three dimensions) whilst the p orbitals are now dumb-bell shaped. The d orbitals look much the same when squared and are not shown.

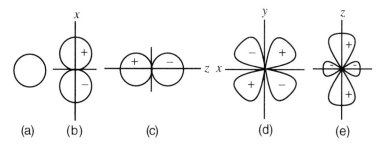

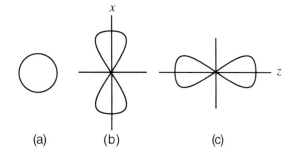

Fig. 1.4 Pictures of the angular part of the wavefunction for selected orbitals; (a) s; (b) p_x; (c) p_z; (d) d_{xy}; (e) d_{z^2}..

Fig. 1.5 Pictures of the square of the angular part of the wavefunction for selected orbitals; (a) s; (b) p_x; (c) p_z..

An important point to remember here is that in the graphs of ψ shown in Fig. 1.4, the signs of the wavefunction on either side of the nodes (important when constructing molecular orbitals) are present as + and –. In the graphs of ψ^2 shown in Fig. 1.5, everything is now positive since this represents the probability of finding an electron. The pictures of orbitals that are most commonly met are derived from those shown in Fig. 1.5 but signs or shading are often included as well, i.e. orbitals are usually drawn containing information from graphs of both ψ and ψ^2. It is important to realise that the + and – signs have <u>nothing</u> to do with charge.

A fuller discussion of orbital types including many excellent pictures can be found in Winter (1993).

1.2.2 The radial part of the wavefunction

The radial part of the wavefunction tells us how the wavefunction varies with distance, r, from the nucleus, i.e. the effective size of the orbital, and we shall see that a square of this function will be particularly useful in understanding many aspects of the structure of the periodic table. Atomic orbitals depend on an exponential function e^{-Br} where B is some constant and r is the distance from the nucleus. This means that ψ falls away exponentially at large r values. For a 1s orbital, a standard expression is $\psi = Ae^{-Br}$ and if we plot this function as a graph we obtain a curve shown in Fig. 1.6. from which we can see that there is a maximum value of ψ at $r = 0$ or at the nucleus. If we now plot a graph of ψ^2 vs. r to get the probability of finding an electron at a particular point, we find that this looks rather similar as shown in Fig. 1.7.

However, we are not so much interested in the probability of finding an electron at a point along a line from the nucleus but rather with the probability of finding the electron on a particular three-dimensional surface. In the case of the 1s orbital being considered here, we can imagine a spherical surface (derived from the angular part) expanding from the nucleus and it will be useful to know the probability of finding an electron at some point on this surface as a function of distance from the nucleus. The greater the distance from the nucleus, the more points on a sphere and so the function we want will be proportional to the surface area of the sphere or $4\pi r^2$. We can therefore plot a probability that an electron is on a surface at a certain

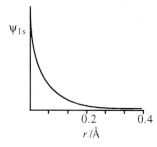

Fig. 1.6 Graph of ψ vs. r for the 1s orbital.

All of the graphs in this chapter should be considered as approximate rather than faithful representations of the mathematical functions from which they are derived.

distance r according to the function $4\pi r^2 . \psi^2$ and a graph of this for the 1s orbital is shown in Fig. 1.8 (a).

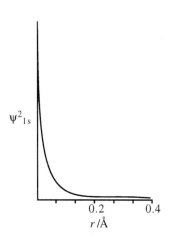

Fig. 1.7 Graph of ψ^2 vs. r for the 1s orbital.

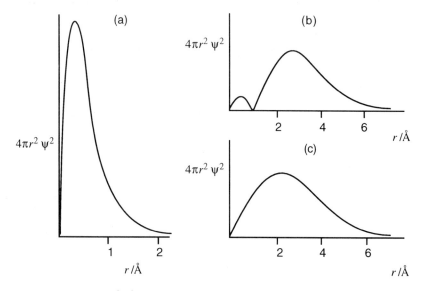

Fig. 1.8 Graphs of $4\pi r^2 . \psi^2$ vs. r for (a) the 1s orbital, (b) the 2s orbital and (c) the 2p orbital drawn approximately to scale.

This function $4\pi r^2 . \psi^2$ is called the *radial probability function* (RPF). When $r = 0$, the RPF is also 0 and so for all orbitals, the function is zero at the centre of the nucleus. We see also that the graph has a maximum, i.e. there is a distance at which we are most likely to find the electron, r_0. For a 1s orbital for hydrogen this is called the Bohr radius and is equal to 0.529 Å. This most probable distance is what is meant by the effective size of the orbital but note that there is a small but finite probability that the electron can be found at large distances from the nucleus as a result of the exponential decay mentioned earlier.

Graphs for the radial probability functions for other hydrogen orbitals are shown in Figs. 1.8 (b), 1.8 (c) and Fig. 1.9 (more accurate, quantitative diagrams are available in most textbooks, see for example, Huheey, Keiter and Keiter) to illustrate some of the more important points. Note, for example, that some graphs have secondary maxima closer to the nucleus and also that the major maximum in the graph of the 2s orbital is slightly further from the nucleus than is the maximum in the 2p orbital, a matter we shall remark upon later, although both are considerably further from the nucleus than that for the 1s orbital.

Some important points for all orbitals which derive from this treatment are as follows:

1. All functions tend to zero at large values of r.
2. Some functions have $4\pi r^2 . \psi^2 = 0$ for certain values of r (other than for $r = 0$) which correspond to nodes (so-called radial nodes). Remember that the total number of nodes for any orbital is given by $n-1$ and since

the number of angular nodes is equal to l, the number of radial nodes is equal to $n-l-1$.

3. All RPFs have a maximum at some value of r, i.e. r_0. If we compare 1s, 2s, and 3s we find that r_0 increases with n. Similarly for 2p vs. 3p vs. 4p, i.e. the value of the principal quantum number n determines the size of the orbital. For a given value of n, the orbitals are of similar size as shown in Figs. 1.8(b) and (c) and in Fig. 1.9, the main maxima being slightly closer to the nucleus the larger the value of l.

4. For orbitals with radial nodes, secondary maxima occur closer to the nucleus. This will be important when we look at polyelectronic atoms and the ordering of orbital energies.

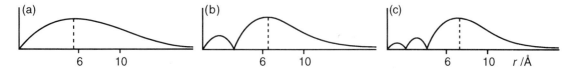

Fig. 1.9 Graphs of RPFs of (a) the 3d orbital, (b) the 3p orbital and (c) the 3s orbital drawn approximately to scale.

It is important to remember that all of the above figures are graphs of mathematical functions and should not be taken as physical representations of how orbitals 'look'. If we try to represent both parts of the wavefunction together we get contour maps or shaded pictures and in the contour plot of the $3p_z$ orbital shown in Fig. 1.10, we see both a radial node and an angular node.

1.3 Energies of orbitals

For an electron in an orbital, the total energy, E, is equal to the sum of the kinetic energy and the potential energy. Both vary depending on position but their sum is fixed. Lower E means more stable, and electrons will go into the lowest energy orbitals available. For hydrogen, E depends *only* on the principal quantum number n and so the 1s orbital is lowest in energy. Therefore in the *ground state* the electron in hydrogen is in a 1s orbital. Hydrogen can also exist in *excited states* where the electron can be in an orbital of higher energy.

The energy level diagram shown in Fig. 1.3 is an illustration of these ideas and shows in particular that E depends only on n. Moreover energies are taken as negative, more negative meaning lower energy and therefore more stable. Zero is defined as the removal of the electron from the atom, i.e. the energy to remove an electron or *ionization energy* is the difference between zero and 1s for the hydrogen atom.

1.4 Polyelectronic atoms

Having considered hydrogen with only a single electron, we are now ready to consider polyelectronic atoms, i.e. the rest of the periodic table. One immediate problem which we encounter is that the Schrödinger equation can only be solved exactly for the hydrogen atom which has one electron (or

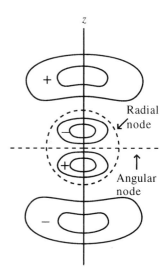

Fig. 1.10 A contour plot of the $3p_z$ orbital.

other one-electron species such as He^+ and Li^{2+}). Solutions to the Schrödinger equation which describe polyelectronic atoms can, in fact, only be solved approximately but calculations indicate that orbitals for other atoms are very similar to those found for hydrogen.

In particular, some important points are as follows:

1. The same quantum numbers are found as for hydrogen.
2. The same angular functions are found so that the orbitals have the same shapes and types, i.e. 1s, 2s, 2p, etc.
3. The radial functions are also similar but firstly they are contracted to smaller radii (i.e. orbitals shrink due to increasing nuclear charge), and secondly, the energies are lower. Most importantly of all, orbital energies now depend on the subsidiary quantum number l as well as upon n for reasons which we shall address shortly.

Since we now have many electrons to deal with and in order to understand the structure of polyelectronic atoms, we must now look and see how orbitals are filled by electrons. There are three rules or principles which we must consider.

1.4.1 The Pauli exclusion principle

The introduction of a fourth quantum number at this point seems a little *ad hoc* but from a more advanced relativistic treatment of the Schrödinger equation, developed by Dirac, all four quantum numbers can be derived directly. We are not concerned here with the reason for the particular numerical values associated with m_S.

In order to understand this, we must first introduce a fourth quantum number. This is labelled m_S and is associated with the electron itself and its *spin*. It can take the values $+\frac{1}{2}$ and $-\frac{1}{2}$, i.e. two distinct values The Pauli principle in its simplest form then states that *no two electrons in the same atom may have the same set of four quantum numbers*. Since each orbital is defined by three quantum numbers, n, l, and m_l, it follows that only two electrons can be associated with any particular orbital, one with $m_S = +\frac{1}{2}$ and the other with $m_S = -\frac{1}{2}$. Two such electrons are said to be *spin paired* or have *opposite spins*.

1.4.2 The *aufbau* principle

This states that electrons go into the lowest energy orbitals available. Thus we build up (hence *aufbau* from German) the electron configuration for an atom by adding electrons, two to each with spins paired, to the lowest energy orbitals. This is represented with the examples shown below which also illustrate the standard form of denoting the electron configuration of an atom.

i.e. H $1s^1$
He $1s^2$
Li $1s^2\,2s^1$
Be $1s^2\,2s^2$
B $1s^2\,2s^2\,2p^1$ (Note that 2p is higher in energy than 2s.)

1.4.3 Hund's first rule

This states that for a set of degenerate orbitals (orbitals with the same energy) electrons will start by going one into each with their spins *aligned* or *parallel*, i.e.

C $1s^2\ 2s^2\ 2p_x^1\ 2p_y^1$
N $1s^2\ 2s^2\ 2p_x^1\ 2p_y^1\ 2p_z^1$
O $1s^2\ 2s^2\ 2p_x^2\ 2p_y^1\ 2p_z^1$etc.

There are two reasons for this. Firstly, electrons go in different orbitals if possible since this places them further apart from each other and thus minimizes the coulombic repulsion resulting from the fact that they have the same charge. Secondly, when two or more orbitals are singly occupied, as for carbon and nitrogen, spin parallel is more stable (for quantum mechanical reasons) than spin paired.

An alternative graphical representation is to use a set of boxes for the orbitals in which arrows denote electrons with the direction indicating the electron spin.

The electronic structure of carbon has been written above as $2p_x^1\ 2p_y^1$ but in terms of p orbital occupation, this is entirely arbitrary; it could just as well have been written as $2p_x^1\ 2p_z^1$.

1.5 Energies of electrons in polyelectronic atoms

The energies of the electrons in non-hydrogen atoms, i.e. atoms with more than one electron, depend on both the quantum numbers *n* and *l*. We will now consider why this is so and shall see that an understanding of this fact is vital to understanding the form of the periodic table.

Let us start by considering some first ionization energies, i.e. the energy required to remove the highest energy or least tightly held electron. For hydrogen, the first ionization energy (1st IE) is 1313 kJ mol^{-1}. For the helium ion, He$^+$, the calculated 1st IE is 5252 kJ mol^{-1} because it is proportional to the square of the nuclear charge, Z^2, and the observed 1st IE is found to be 5250 kJ mol^{-1}, i.e. good agreement since the He$^+$ ion has one electron like the hydrogen atom.

However, for the helium atom, He, with two electrons, the observed 1st IE is found to be much less at 2372 kJ mol^{-1}. Clearly the two electrons in the 1s orbital of the He atom are at a higher energy than the single electron of He$^+$ which arises, in part, because of electron-electron repulsion which partly offsets the electron-nucleus attraction. It is as if the electrons partly shield each other from the full attraction of the nucleus and for He, each electron feels only about $1\frac{1}{3}$ of the +2 charge of the nucleus and there is a therefore a shielding effect of about $\frac{2}{3}$ of an electron; incomplete but effective shielding of the electrons, by each other, from the positive charge of the nucleus.

For lithium, Li, the observed 1st IE is 520 kJ mol^{-1}, which is quite low. There is therefore a considerable shielding of the outermost electron in Li from the +3 nuclear charge and the *effective nuclear charge* attracting this 2s electron of Li is only about $+1\frac{1}{4}$ instead of +3. The outer 2s electron is therefore very effectively shielded from the nucleus by the two 1s electrons and we can understand this in the following way.

If we look at the radial probability functions for the 1s and 2s orbitals superimposed on the same graph (see Fig. 1.11), we can see that most of the 2s orbital lies well outside the main part of the 1s electron density, i.e. the maximum for 2s is at a considerably greater distance from the nucleus than the maximum for 1s. The effect of this is that, in the case of Li, the two 1s electrons shield the 2s electron very effectively from the nuclear charge

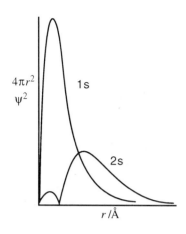

$4\pi r^2$
ψ^2

1s

2s

$r\,/\text{Å}$

Fig. 1.11 Graphs of RPFs for the 1s and 2s orbitals superimposed.

although they provide much less shielding for each other. Moreover, the electron in the 2s orbital does not really shield the 1s electrons at all.

This leads us to consider the electrons in an atom in two groups. The *valence electrons*, are those with the highest energies (normally those with the highest value of *n*) and are involved directly with chemistry and bonding, whilst the *core electrons*, with lower values of *n*, are in completely filled orbital groups and act merely as a less-than-complete shield of the nuclear charge. The latter are too tightly bound to the nucleus to be directly involved in chemistry, i.e. their energy is too low, but nevertheless play an important rôle as a result of their shielding influence on the valence electrons. In fact, the very existence of a periodic pattern in the properties of the elements implies that it is only the valence electrons which are responsible for their chemical behaviour. It is therefore appropriate to think of the nucleus and the core electrons as providing an effective charge or potential (pseudo potential as it is sometimes called) in which the valence electrons move. This concept is very useful in simplifying quantum chemical calculations.

Accurate wavefunctions are available from computer programs for all known atoms and so are all the orbital energies and these agree well with the observed ionization energies. What we as chemists like, however, is a simple model with as few numbers to remember as possible. Slater has suggested a set of simple rules to estimate shielding effects and hence to calculate effective nuclear charge values, which can then be used to calculate approximate orbital energies. We will look at these in some detail in the following chapter.

In discussing the element lithium, we placed the outermost electron in the 2s rather than the 2p orbital and at this stage we need to consider why it is that for Li, and all polyelectronic atoms, 2s is lower in energy than 2p. In other words why do orbital energies now depend on *n* and *l* whereas for hydrogen they depend only upon *n*. The answer can be found by again considering the radial parts of the wavefunctions, and in the particular importance of the small inner lobes seen in some of the radial probability functions.

For Li there are two 1s electrons and a third electron to be accommodated in an orbital of higher energy. Why does this go into 2s rather than the 2p? As we have seen, when considering the low ionization energy of Li, the two 1s electrons partly shield the third electron by getting between it and the nucleus since the 1s electrons lie, on average, much closer to the nucleus (Fig. 1.11). However, a small but significant part of the 2s orbital lies even closer to the nucleus than the maximum in the 1s lobe. This is associated with the minor maximum in the graph for 2s which results from the presence of a radial node. It is this contribution which lowers the energy of 2s relative to 2p since the 2p orbital contains no radial nodes and hence no minor maximum close to the nucleus (Fig. 1.12). The electron in the 2s orbital can therefore be said to *penetrate* the core (i.e. 1s) better. The effect may seem to be small when looking at the graphs especially when it is noted that the major maximum for 2s is actually further from the nucleus, albeit only slightly, than the maximum for 2p. However, calculations of the potential energies of electrons indicate that this small penetration is all-important in determining the ordering of the orbital energies and thus 2s lies lower in energy than 2p and it is the 2s orbital which is occupied first.

Similar arguments can be offered to explain why 3s penetrates more than 3p which, in turn, penetrates more than 3d and therefore why the orbitals are filled in this order.

One point we should note at this stage is that we must be careful in using graphs of the radial probability functions for hydrogen atoms to explain properties of polyelectronic atoms. As we shall see later, graphs for a particular orbital can change markedly, with respect to the distance of the maxima from the nucleus, depending on whether other orbitals are filled or not. In the case of lithium and the 2s and 2p orbitals described above, the discussion based around the graphs in Fig. 1.12 is largely accurate because the radial extensions of the 2s and 2p orbitals are still similar (as they are in hydrogen) even though 1s is filled. For the heavier elements, this is no longer the case and it is found that s orbitals have increasingly smaller radial extensions relative to p orbitals with the same value of n as n increases. This has important consequences for bonding as we shall see later.

We are now in a position to look at why it is that the periodic table has the form it does. After H and He, the 2s orbital is filled (Li and Be) and then the three 2p orbitals giving a total of six elements (B through Ne). This completes the filling of the orbitals of the second quantum shell. For the next element, sodium, the outermost electron goes into the 3s orbital and the pattern of the previous shell repeats itself with the filling of 3s and 3p taking us as far as argon. It is apparent from an inspection of the periodic table, however, that the outermost electron for the next element, potassium, resides not in one of the 3d orbitals, as might have been expected, but rather in the 4s orbital. So far we have observed that orbital energies are dependent primarily on n with a secondary dependence on l. Clearly this is an over-simplification as an orbital with $n = 4$ is being filled before all of the $n = 3$ orbitals have been filled.

We can understand the reason for this fact by considering the graphs shown in Fig. 1.13 which show the radial probability functions for the 4s and 3d orbitals and the extent to which they penetrate the 3s/3p core.

It is apparent that whilst 3d has its maximum closer to the nucleus than does 4s, the inner lobes of 4s penetrate the core very effectively. Calculations show that, with 3s and 3p filled, an electron in 4s is less efficiently shielded than an electron in 3d and therefore feels a greater effective nuclear charge and so is lower in energy (the fact that 3d and 4s have similar radial extensions means that effective nuclear charge is the dominant factor). The outermost electrons for potassium and calcium therefore reside in 4s. Note, however, that once 4s has been filled, it shields 3d very poorly since much of the electron density in 4s lies outside the maximum of 3d. The 3d orbitals therefore fall sharply in energy as soon as 4s is filled and these five orbitals are the next to be filled which gives rise to the ten elements of the first transition series, Sc through Zn. The next lowest energy orbitals available are the three 4p orbitals and these are filled next with the elements Ga through Kr. The rest of the periodic table can be built up and understood in much the same way. The situation gets a little complicated with the later

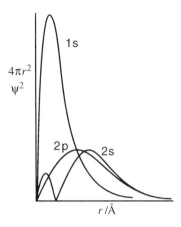

Fig. 1.12 Graphs of the RPFs for the 1s, 2s and 2p orbitals superimposed.

A periodic table is shown on the inside of the back cover.

An indication of the difference in radial extension of the nd and $n+1$s orbitals is afforded by the following data for the elements Cr, Mo and W.

	r_{max} nd (pm)	r_{max} $n+1$s (pm)
Cr	46	161
Mo	74	168
W	79	147

transition elements and the lanthanides because there are many orbitals that are very close in energy but we will not be concerned here with this part of the periodic table.

We must exercise care in using graphs like Fig. 1.13. In Figs. 1.8 and Fig. 1.9, the graphs of the 3s, 3p and 3d orbitals shown are those for the hydrogen atom which contains only one electron and for which shielding and penetration effects are therefore irrelevant. With polyelectronic atoms the radial probability plots will change considerably as a result of these shielding effects. Thus, whilst for hydrogen, the 4s orbital has a maximum much further from the nucleus than that of 3d, for elements like potassium, these two orbitals are much the same size.

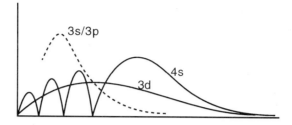

Fig. 1.13 Graphs of the RPFs for the 4s and 3d orbitals superimposed on the 3s/ 3p core (dashed line).

An important point which derives from this discussion is that the energies of orbitals (or, to be more precise, the order in which they are filled) are very dependent, as a result of shielding and penetration effects, on whether or not other orbitals are filled or unfilled. A very useful graph of the variation in the orbital energies for the elements is shown in Fig. 1.14 and we will refer to this again in subsequent chapters.

In Fig. 1.14, the 3d orbital is drawn at higher energy than the 4s for elements with an atomic number of around 20–25. Whether or not the 3d orbital ever does rise above 4s in energy is a matter of some debate, but the graph and associated discussion are adequate for our purposes. In any event, we must be careful to distinguish between statements about orbital energies and about the order in which orbitals are filled since orbital energies vary depending on whether or not other orbitals are filled. This matter is addressed again in the next chapter.

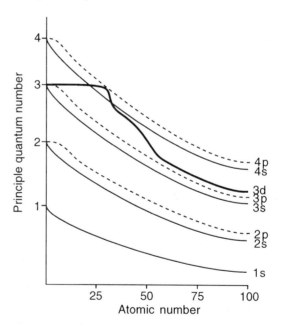

Fig. 1.14 Approximate orbital energies up to 4p for the elements up to Z = 100.

From Fig. 1.14 we can see that for hydrogen we have the special simple case that E depends only on n and so 2s and 2p are degenerate as are 3s, 3p and 3d.

For all other atoms the s, p and d orbitals for any given n are *not* degenerate due to differences in penetration and shielding effects. All s orbitals penetrate the inner shells well so they feel a higher effective nuclear charge than corresponding p and d orbitals; they all drop steadily as the nuclear charge increases (solid lines).

All p orbitals also penetrate the inner shells well except that they are well shielded from the nucleus by the 1s electrons so they drop steadily after staying nearly level for the first few elements. Thus p orbitals are a little above s orbitals (dashed lines). 1s, 2s, 2p, 3s, 3p maintain the same order for all elements.

Returning to 3d orbitals, note that once they are filled they drop quickly in energy (bold line) and approach 3s and 3p since they are very poorly shielded by the next orbitals being filled, i.e. 4p. Thus for Ga and beyond, 3d are now lower in energy than 4s and act as core rather than valence electrons. Therefore the electron configuration of Ga is [Ar] $3d^{10}$ $4s^2$ $4p^1$ and it resembles Al, [Ne] $3s^2$ $3p^1$, in its chemistry. The d orbitals do play a part in Ga chemistry but only a minor one; we will look at this in more detail in later chapters.

Two final points are important, regarding Fig. 1.14.

1   Note that for core orbitals, the individual s, p and d orbitals with the same value of n come close together in energy; i.e. they become more hydrogen-like.
2   All of the above discussion is for neutral atoms only; for ions the ordering can and does change.

The second point above regarding neutral atoms vs. ions requires a little more explanation. As we have seen, potassium has the valence electronic configuration $4s^1$ and scandium, the first element of the transition series, is $4s^2$ $3d^1$. This is generally true for the first row transition elements in that the neutral atoms have the electronic configuration $4s^2$ $3d^n$ (except for Cr which is $4s^1$ $3d^5$ and Cu which is $4s^1$ $3d^{10}$). In discussing the cations of the transition elements, however, it is found that the 3d orbitals are lower in energy than the 4s orbital (i.e. are filled first) and the valence electron configuration of the metal +2 and +3 ions are given as $3d^n$. This is often confusing in that it seems that whilst electrons are added to 4s and then 3d for the atoms, they are removed in the reverse order to form the ions. To resolve this difficulty, we must recognize that the relative ordering of the orbital energies is also dependent on the charge of the atom or ion. Thus, whilst 4s is lower than 3d for the neutral atoms, the reverse is true for the +2 and +3 ions. The reason for this is that for the ions, the effective nuclear charge felt by the outer electrons is much larger with the result that all orbitals become contracted. Penetration effects therefore become less important and orbitals become more hydrogen-like such that energies become more dependent on n rather than l.

In conclusion, we now have a fairly good rationalization for the structure or form of the periodic table and in the next chapter we will go on to look at the periodic trends in various atomic properties.

2 Periodicity in the properties of s- and p-block atoms

In this chapter we will look at a range of properties of isolated atoms of the s- and p-block elements, which will introduce us to the ideas of periodicity, and which will be useful later when we consider both the elements themselves and the compounds they form.

2.1 Effective nuclear charge

In the previous chapter we introduced the concept of *effective nuclear charge* when considering the energies of the valence electrons and from this it was clear that the outermost electrons in an atom feel a nuclear charge which is far less than the actual nuclear charge due to the shielding effects of other electrons. We shall see that this idea of effective nuclear charge is useful in understanding many aspects of periodicity, but in order to utilize and appreciate the concept fully, a semi-quantitative scale is desirable. There are several scales which have been devised, but one of the most widely used, because of its simplicity, is a scheme devised by Slater and known, not surprisingly, as *Slater's rules*.

The object of Slater's rules is to estimate, for a particular electron in an atom (or ion), the strength of the shielding effect of the other electrons present, and from this to calculate a shielding constant, S. This value can then be used to calculate the effective nuclear charge, Z^*, according to the following equation, where Z is the actual nuclear charge:

$$Z^* = Z - S$$

To calculate the shielding constant for a particular ns or np electron (Slater's rules makes no distinction between them), the following rules apply:

For d and f electrons, the rules are slightly different but we will not be concerned with the d- and f-block elements here.

(1) electrons with a higher n contribute zero, i.e. no shielding;
(2) electrons with the same value of n contribute 0.35, i.e. not very good shielding;
(3) electrons with a value of n one less than our chosen electron contribute 0.85, i.e. rather better shielding;
(4) electrons with lower values of n contribute 1.00, i.e. complete shielding.

A couple of examples will illustrate these ideas. For lithium, the electron configuration is $1s^2 2s^1$. Thus for the 2s electron, the shielding constant is $2 \times 0.85 = 1.70$. Hence Z^* for the Li 2s electron $= Z - S = 3 - 1.70 = 1.30$. In the case of nitrogen, the electron configuration is $1s^2 2s^2 2p^3$. Therefore S for one of the 2p electrons is 4×0.35 (i.e. the two remaining 2p electrons and the 2s pair) = 1.40, plus 2 × 0.85 for the 1s pair giving a total of 3.10. Hence $Z^* = 7 - 3.10 = 3.90$.

Values of Z^* for the s- and p-block elements calculated from Slater's rules are given in Table 2.1 and a particularly important trend to note is that the effective nuclear charge increases quite substantially on crossing a period from left to right, the importance of which we shall see later. Values increase to a much smaller extent on descending a group and, because of how the values are calculated according to the rules, become constant lower down.

Table 2.1 Calculated values of Z^* from Slater's rules for the valence n s or n p electrons

Li	Be		B	C	N	O	F	Ne
1.30	1.95		2.60	3.25	3.90	4.55	5.20	5.85
Na	Mg		Al	Si	P	S	Cl	Ar
2.20	2.85		3.50	4.15	4.80	5.45	6.10	6.75
K	Ca		Ga	Ge	As	Se	Br	Kr
2.20	2.85		5.00	5.65	6.30	6.95	7.60	8.25
Rb	Sr		In	Sn	Sb	Te	I	Xe
2.20	2.85		5.00	5.65	6.30	6.95	7.60	8.25
Cs	Ba		Tl	Pb	Bi	Po	At	Rn
2.20	2.85		5.00	5.65	6.30	6.95	7.60	8.25

In considering the values in Table 2.1, it should again be stressed that Slater's rules are very approximate and do not take into account such factors as the difference in the extent to which electrons in s and p orbitals penetrate the core electron density. Other rules, such as those devised by Clementi and Raimondi [see Huheey, Keiter and Keiter (1993)] explicitly deal with such effects but at the expense of simplicity.

2.2 Ionization energies

The ionization energy (sometimes called ionization potential or electron binding energy) is the energy required to completely remove an electron from an atom in the gas phase. For example, Eqn. 2.1 illustrates the first ionization energy of a neutral atom, E.

$$E (g) \quad \rightarrow \quad E^+ (g) + e^- \qquad (2.1)$$

The second ionization energy will be the energy required to remove an electron from the singly charged cation E^+ according to Eqn. 2.2. In all cases, the second ionization energy will be greater than the first since it is more difficult to remove an electron from a positively charged ion than from a neutral atom.

$$E^+ (g) \quad \rightarrow \quad E^{2+} (g) + e^- \qquad (2.2)$$

Let us look first at the overall trends in first ionization energies of the s- and p-block elements, which are presented in Table 2.2. In global terms, it is apparent from Table 2.2, and also from the graph shown in Fig. 2.1, that there is a general tendency for ionization energies to increase as we cross the periodic table from left to right and to decrease as a group is descended. Moreover, a sharp drop occurs when we start a new row (i.e. go to the next higher value of n). The trend across a period is to be expected on the basis of the effective nuclear charges discussed in section 2.1 since, as Z^* increases across the period, the electrons will be more tightly held and therefore require more energy to remove although it is clear from Fig. 2.1 that the trend in ionization energies is not linear as might be anticipated from Slater's rules alone (see below). The sharp drop which occurs on starting a new row is also to be expected on the basis of Z^*, but the trend down a group is perhaps unexpected since values for Z^* increase rather than decrease down a group, before becoming constant. In fact, the most important factor in a group is not so much the charge the electron feels, but rather its distance from the nucleus. Thus, we should recall from Chapter 1 that the maximum in the radial probability function graph for an orbital is very dependent on the value of the principle quantum number n, the average distance from the nucleus generally increasing substantially with increasing n. Thus, even though the effective nuclear charge is increasing down a group, this is outweighed by the fact that the electron is further away from the nucleus and is therefore easier to remove.

Table 2.2 First ionization energies (kJ mol^{-1}) for the s- and p-block elements

Values in Table 2.2 are taken from Emsley (1989).

H 1312							He 2372
Li 513	Be 899	B 801	C 1086	N 1402	O 1314	F 1681	Ne 2081
Na 496	Mg 738	Al 578	Si 786	P 1012	S 1000	Cl 1251	Ar 1520
K 419	Ca 590	Ga 579	Ge 762	As 947	Se 941	Br 1140	Kr 1351
Rb 403	Sr 549	In 558	Sn 709	Sb 834	Te 869	I 1008	Xe 1170
Cs 376	Ba 503	Tl 589	Pb 715	Bi 703	Po 812	At 930	Rn 1037

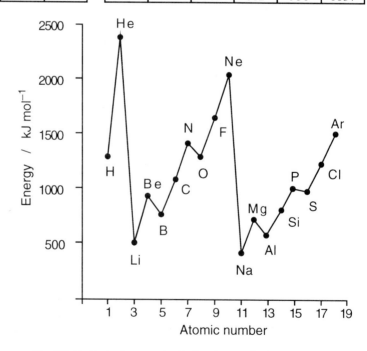

Fig. 2.1 First ionization energies for the elements hydrogen to argon.

If we now look at Fig. 2.1 in more detail, we can see that whilst the ionization energies for Li to Ne generally increase, there are kinks in the graph at boron and oxygen. In the former case, boron has a lower first ionization energy than beryllium because the single 2p electron in boron is quite effectively shielded by the 2s pair and is therefore more readily lost (see Fig. 1.12 and note the minor maximum close to the nucleus for the 2s orbital), an aspect of shielding not taken into account in Slater's rules. In the latter case, oxygen has a lower ionization energy than nitrogen since, whereas nitrogen has one electron in each of the three p orbitals, electrons have to pair up for oxygen and one p orbital becomes doubly occupied. Two electrons in one orbital are confined to the same region of space and thus experience a greater coulombic repulsion and it is this which accounts for the lower than expected first ionization energy of oxygen.

After neon there is a sharp drop on moving to sodium, as expected on the basis of the discussion above, since Z^* values for Ne and Na are estimated to be 5.85 and 2.20 respectively, and also we have moved from the second quantum shell to the third and therefore the outermost electron in sodium is now much further from the nucleus. The subsequent trend from sodium to argon is very similar to that seen for lithium to neon, and for exactly the same reasons.

Let us now look in more detail at trends within a group which we can illustrate by considering groups 13 and 18. In group 18 (and in groups 1 and 2) there is a regular decrease in the first ionization energy as the group is descended due to the increase in the principal quantum number associated with the outermost electrons, and the resulting increase in the distance of this electron from the nucleus. As mentioned above, this factor is more important than the small increase in Z^* as the group is descended.

If we look at group 13, however, there at first seems to be little regularity in the ionization energies. Aluminium is lower than boron as we would expect, but the value for gallium is actually slightly greater than that for aluminium. The obvious difference between B vs. Al and Al vs. Ga is that Ga is preceded by the first row of the d-block elements and it is to this factor that we should look for an explanation. Thus, in the absence of the 3d row, gallium would come after calcium and we would expect a regular trend from B to Al to Ga. With the inclusion of the elements Sc through Zn, however, we have added ten units of positive charge to the nucleus but the corresponding ten electrons are in 3d orbitals which do not provide very effective shielding for the 4p electrons since 4p, with its minor maxima close to the nucleus (like 4s), pentrates 3d quite well. As a result, the effective nuclear charge felt by the 4p electron in gallium is larger than it would be in the absence of the 3d row, thereby making it harder to remove. Slater's rules as they are given in Table 2.1 are not particularly helpful in this regard since, although Z^* for Ga is larger than that for Al, the same is true for Al vs. B. Nevertheless, we can use these rules as a guide in calculating the effect as the value of Z^* for Ga in the absence of the 3d row, i.e. if Ga were [Ar] $4s^2\ 4p^1$, Z^* would be 3.50 rather than the value given in Table 2.1 of 5.00. Clearly, the increase in Z^* more than offsets the increase in n in determining the ionization energy in this case.

A complete periodic table with the group numbering system is presented on the inside of the back cover.

Note that the ionization energies for the series of elements B, Al, Sc, Y, and La show a much more regular decrease since there are no disparate shielding effects resulting from different core electron configurations.

On moving further down the group, we see that the ionization energy for indium is less than gallium, the general trend we would expect, but at thallium, a substantial increase occurs such that the first ionization energy of thallium is larger than for all the elements in this group except boron! Two factors are important here. Firstly, the inclusion of the 4f row will lead to an increase in the effective nuclear charge felt by a thallium 6p electron for exactly the same reasons as those given above for the effect of the 3d row on a 4p electron. Secondly, so-called relativistic effects, to be described later, result in a lowering of the energy of the 6p orbital.

The factors described above, namely increasing effective nuclear charge as a result of d and f rows, and relativistic effects, will have important consequences for the chemistry of the 4p and 6p elements as we shall see, although they are most pronounced in terms of ionization energies for the elements of group 13. Nevertheless, we can also see the effects in group 14. Note that the ionization energy of germanium is very similar to that of silicon and that lead has a higher first ionization energy than tin. Moreover, these trends are continued in groups 15 and 16 where the similarity in ionization energies between the 3p and 4p and between the 5p and 6p elements is apparent.

We will return to ionization energies later when discussing the chemistry of the elements where trends in the sums of particular ionization energies will be important.

2.3 Electron affinities

By convention, electron affinities are given a *positive* value when energy is given out (note that this corresponds to a *negative enthalpy* of electron attachment).

The electron affinity of an atom is defined as the energy change which occurs when an electron is added to an atom (or ion) according to Eqn 2.3; it is the reverse of ionization. Experimental values for the electron affinities of the s- and p-block elements are presented in Table 2.3 and Fig. 2.2 from which a number of trends are apparent.

$$E\ (g)\ +\ e^- \quad \rightarrow \quad E^-\ (g) \tag{2.3}$$

Table 2.3 Electron affinities ($kJ\ mol^{-1}$) for the s- and p-block elements

Values in Table 2.3 are taken from Emsley (1989).

H							He
73							<0
Li	Be	B	C	N	O	F	Ne
60	−18	23	122	-7	141	322	<0
Na	Mg	Al	Si	P	S	Cl	Ar
53	−21	44	134	72	200	349	<0
K	Ca	Ga	Ge	As	Se	Br	Kr
44	−186	36	116	77	195	324	<0
Rb	Sr	In	Sn	Sb	Te	I	Xe
47	−146	34	121	101	190	295	<0
Cs	Ba	Tl	Pb	Bi	Po	At	Rn
45	−46	30	35	101	186	270	<0

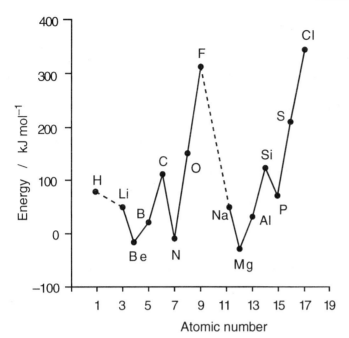

Fig. 2.2 Electron affinities for the elements of the first two short periods with the exception of the noble gases.

We can note first of all that, with the exception of the noble gases, electron affinities generally increase (i.e. electron attachment becomes more exothermic) on crossing the p-block periods from left to right. This is the same trend as seen for ionization energies and for much the same reason. Thus, as the effective nuclear charge increases, the energy released on adding an electron also increases. However, the trend is not regular most notably for the group 15 elements which appear to have lower electron affinities than might be expected. An explanation can be found by considering the electron configurations of the neutral atoms which are np^3. Thus, an extra electron must be added to an orbital that is already half filled which results in increased coulombic repulsion; exactly the same argument was used to account for the lower than expected ionization energy of oxygen compared to nitrogen.

In the case of the s-block elements, we should note two points. Firstly, the electron affinities for the group 1 elements are larger than those for the corresponding group 13 elements, and secondly, that the values for the group 2 elements are all negative (i.e. endothermic). An explanation of the former observation lies in the fact that formation of a group 1 anion results in a filled s orbital which has a certain stability, whereas in the second case, addition of an electron to a group 2 element places an electron in a higher energy p orbital which is well shielded by the preceding ns^2 pair (c.f. the explanation of why boron has a lower first ionization energy than beryllium).

In terms of group trends, we should note that the electron affinities for the 3p and 4p elements in the same group are similar as are those for the 5p and 6p elements (lead would seem to be an anomaly here). This reflects the larger

The situation for the noble gases is a little different in the case of electron affinities compared with ionization energies. The electron affinities for the noble gases are all calculated to be endothermic (-30 to -40 kJ mol^{-1}, quoted as <0 in Table 2.3). Thus, although an outer electron in a group 18 element feels a high effective nuclear charge, an extra electron must occupy an orbital of the next principal quantum shell which is further from the nucleus, and it will also be well shielded by the preceding s^2p^6 core.

effective nuclear charges found for the 4p and 6p elements as a result of the filling of the 3d row and the 4f row respectively, a factor we referred to in section 2.1 above when considering a similar effect with ionization energies. This offsets the anticipated trend to lower values as the groups are descended.

Finally we can note that the electron affinities for the first row elements are generally lower than those for the second row elements, particularly in the case of groups 15, 16 and 17. This is a result of the very small sizes of the atoms of the first row elements, i.e. N, O and F (see later), which is important since the addition of an electron leads to a small negative ion with a high charge-to-size ratio resulting in large coulombic or interelectron repulsions. It is largely this size effect which leads to the first row elements having lower than expected electron affinities.

2.4 Covalent and ionic radii

The concept of atomic radii whether they be covalent, ionic or metallic must be considered carefully since we know from the discussion in Chapter 1 on atomic orbitals, that an atom is not a hard sphere and that the electron density drops away exponentially as the distance from the nucleus increases. Nevertheless, such radii are particularly useful and we shall look briefly in this section at some of the observed trends in covalent and ionic radii.

Strictly speaking, a covalent radius is not a property of an isolated atom.

A covalent radius is defined as half the length of a symmetrical, homonuclear element-element (E–E) bond (usually a single bond although we can calculate multiple bond radii as well), i.e. half of the E–E distance. It is particularly useful to have such radii available since we can use them to calculate distances for previously unknown bonds. Thus, for example, if we calculate the covalent radii for A from A–A and that for B from B–B we should be able to estimate the bond length A–B. In general this approach works well although we must be careful to take into account any electronegativity difference between the elements (see later) since any degree of ionicity or charge separation will lead to a shortening of the bond. A–B bond distances can be calculated from covalent radii more accurately by using the Shomaker-Stevenson relationship which explicitly corrects for any electronegativity differences between A and B.

For example, a value for the covalent radius of fluorine of 0.54 Å rather than 0.64 Å has recently been suggested from a reappraisal of the bond lengths of a number of fluorine compounds.

Covalent radii for the s- and p-block elements are presented in Table 2.4 but it is important to keep in mind that the values quoted for covalent radii sometimes differ depending on the source consulted because their calculation is not always as straightforward and unambiguous as might be thought. However, it is general trends in which we are interested and those are clearly apparent from the values given in Table 2.4.

Two obvious trends may be discerned. Firstly, there is a decrease in covalent radius on moving from left to right across a period. Since all electrons are being added to the same principal quantum shell we might expect, other things being equal, that the atoms in any particular period would remain the same size. Of course, other things are not equal, most importantly, the effective nuclear charge which is increasing. The electrons are therefore more tightly held, the size of the orbital is contracted, and so the covalent radii are smaller.

Table 2.4 Covalent radii (Å) for the s- and p-block elements

H							He
0.30							—
Li	Be	B	C	N	O	F	Ne
1.23	0.89	0.88	0.77	0.70	0.66	0.64	—
Na	Mg	Al	Si	P	S	Cl	Ar
1.54	1.36	1.25	1.17	1.10	1.04	0.99	—
K	Ca	Ga	Ge	As	Se	Br	Kr
2.03	1.74	1.25	1.22	1.21	1.17	1.14	1.10
Rb	Sr	In	Sn	Sb	Te	I	Xe
—	1.92	1.50	1.40	1.41	1.37	1.33	1.30
Cs	Ba	Tl	Pb	Bi	Po	At	Rn
2.35	1.98	1.55	1.54	1.52	1.53	—	—

Values in Table 2.4 are taken from Emsley (1989). Covalent radii are not available for some elements.

Secondly, there is a trend to increasing covalent radius as the groups are descended. This effect results from the fact that electrons are being placed in orbitals with increasing principal quantum number and therefore lie further from the nucleus. As with ionization energies and electron affinities, it is this factor which is of primary importance rather than any increase in effective nuclear charge derived from Slater's rules.

Another point to note is the close similarity in size between the 3p and 4p elements. This is another consequence of the higher effective nuclear charge for the 4p elements resulting from the presence of the 3d row which serves to contract the orbitals and therefore make the elements smaller (note that gallium is the same size as aluminium). This is sometimes referred to as the 'scandide' contraction (from scandium) and is analogous to the better-known lanthanide contraction which results in the 5d elements being very similar in size to the 4d elements due to the presence of the 4f elements which precede the 5d row.

Ionic radii show much the same trends as covalent radii and we will not dwell on the matter in much detail except to show that for the well-known halide anions (shown to the right), the radii increase regularly as the group is descended. Note that the ionic radii for the halides are significantly larger than the covalent radii shown in Table 2.4. We should expect that radii will increase with increasing negative charge (or decreasing positive charge) due to increasing inter-electron repulsions, and a good illustration of this effect is seen by comparing the ionic and covalent radii for germanium and tin shown below (values chosen for ionic radii are very dependent on the coordination number; the value chosen here is 6).

F^-, 1.33 Cl^-, 1.81
Br^-, 1.96 I^-, 2.20
(values in Å for 6-coordination)

	Radius (Å)		
	Covalent	+2 ionic	+4 ionic
Ge	1.22	0.87	0.67
Sn	1.40	1.12	0.83

A graph of covalent radii for the elements of the first two periods (excluding He and Ne) and hydrogen is shown in Fig. 2.3.

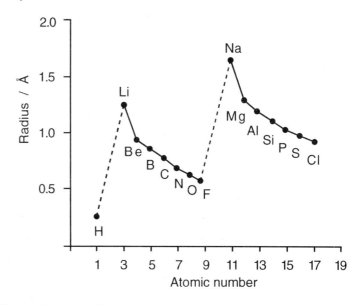

Fig. 2.3 Covalent radii for hydrogen and the elements of the first two short periods.

2.5 Electronegativity

It is said by some that electronegativity is a rather artificial concept with the clear implication that its value is somewhat limited. While it may be true that element electronegativities are not amenable to direct experimental measurement (although see later), they are no more artificial or nebulous than the covalent and ionic radii used extensively by all structural chemists. Moreover, as will be seen in the following discussion and in later chapters, electronegativity can be used as an extremely powerful organizing principle for understanding the nature of the elements themselves and the types of compound they form with each other. It has even been argued that electronegativity should be considered as the third dimension of the periodic table.

The concept of electronegativity (hereafter referred to with the symbol χ) was first recognised by Berzelius more than 150 years ago, but it was Pauling in the 1930s who first provided a quantitative scale together with a simple and concise definition as *the ability of an atom to attract electron density towards itself in a molecule*. The Pauling scale of electronegativity, traditionally quoted on a scale of 0 to 4, is still one of the most widely used by chemists and a compilation of modified Pauling electronegativity values for the s- and p-block elements is given in Table 2.5. Other commonly used scales are those of Allred and Rochow, in which the element electronegativity is defined as proportional to Z^*/r^2 (where r is the covalent radius), and the Mulliken scale where χ is proportional to half the sum of the first ionization energy and the first electron affinity.

Table 2.5 Pauling electronegativities for the s- and p-block elements

H 2.20								He —
Li 0.98	Be 1.57	B 2.04	C 2.55	N 3.04	O 3.44	F 3.98	Ne —
Na 0.93	Mg 1.31	Al 1.61	Si 1.90	P 2.19	S 2.58	Cl 3.16	Ar —
K 0.82	Ca 1.00	Ga 1.81	Ge 2.01	As 2.18	Se 2.55	Br 2.96	Kr —
Rb 0.82	Sr 0.95	In 1.78	Sn 1.96	Sb 2.05	Te 2.10	I 2.66	Xe —
Cs 0.79	Ba 0.89	Tl 2.04	Pb 2.33	Bi 2.02	Po (2.0)	At (2.2)	Rn —

Values in Table 2.4 are taken from Emsley (1989).

This is not the place to discuss electronegativity in great detail, particularly the merits of the many different scales which have been developed over the years, but it will be instructive to comment on the recent work of Allen in this area (Allen, *J. Am. Chem. Soc.*, (1989), **111**, 9003). It is Allen's contention that the electronegativity of an element is *the average one-electron energy of valence shell electrons in the ground state free atoms* and that it can be calculated from experimental spectroscopic data according to the Eqn 2.4 where ε_p and ε_s are the p and s ionization energies and m and n the number of p and s electrons. Electronegativity values calculated according to this formula, χ_{spec}, and adjusted to the Pauling scale are presented in Table 2.6.

$$\chi_{spec} = (m\,\varepsilon_p + n\,\varepsilon_s)\,/\,(m + n) \qquad (2.4)$$

Table 2.6 Allen electronegativities calculated for the s- and p-block elements

H 2.300								He 4.157
Li 0.912	Be 1.576	B 2.051	C 2.544	N 3.066	O 3.610	F 4.193	Ne 4.787
Na 0.869	Mg 1.293	Al 1.613	Si 1.916	P 2.253	S 2.589	Cl 2.869	Ar 3.242
K 0.734	Ca 1.034	Ga 1.756	Ge 1.994	As 2.211	Se 2.424	Br 2.685	Kr 2.966
Rb 0.706	Sr 0.963	In 1.656	Sn 1.824	Sb 1.984	Te 2.158	I 2.359	Xe 2.582

Allen electronegativities are not yet available for the 6s and 6p elements.

By and large, the Pauling values, χ_P, and χ_{spec} are rather similar and we will not look at any comparisons in detail except to note that whereas the Pauling scale indicates that chlorine is more electronegative than nitrogen, the Allen scale suggests that the reverse is true. The relative electronegativities of these two elements has been a longstanding matter of uncertainty in inorganic chemistry.

Allen has also made the point referred to earlier that we can think of electronegativity as being the third dimension of the periodic table and this is shown graphically in Fig. 2.4.

Pauling electronegativity values have been used to produce Fig. 2.4 in order that the 6s and 6p elements might be included.

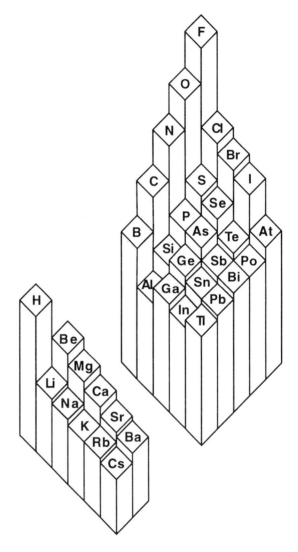

Fig. 2.4 A view of the periodic table for the s- and p-block elements showing electronegativity as a third dimension.

This three-dimensional view allows us to see immediately the trends in electronegativity for the s- and p-block elements as a whole. Thus, if we look at the trends by row, as shown specifically in the graph in Fig. 2.5, it is apparent that electronegativity increases fairly regularly across the periods with the rate of increase tending to decrease as the s- and p-blocks are descended.

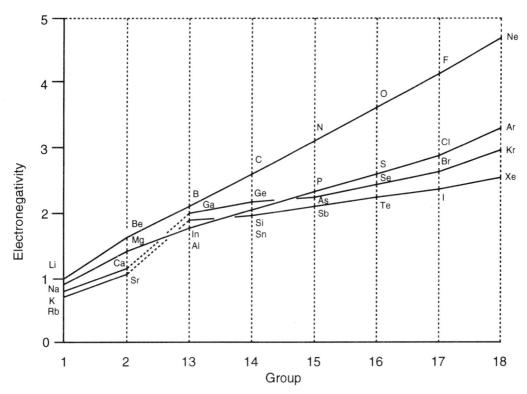

Fig. 2.5 Allen electronegativities as a function of row.

In Fig. 2.6, a graph is shown of Allen electronegativities according to group from which a number of other trends are apparent. Firstly, there is a general decrease in electronegativity as the groups are descended, but the similarity in the values for the 3p and 4p elements is clear. Moreover, in the case of groups 13 and 14, the electronegativity of the 4p element is actually greater than that of the 3p element, i.e. Ga > Al and Ge > Si. This is a further example of the consquences of the increased effective nuclear charges of the 4p elements resulting from the presence of the 3d row.

We shall have more to say on the electronegativities of the elements in the next chapter when we look at the properties of the elements themselves and where the meaning of the metalloid band in Fig. 2.6 will be discussed. Remember, however, that the values of χ_{spec} discussed above are for neutral, isolated atoms. In compounds of the elements, the precise values will depend on a number of factors such as oxidation state and hybridization. We will look at this in more detail later, but it is known, for example, that the electronegativity of thallium is considerably larger in thallium(III) complexes ($\chi_P = 2.04$) than in those of thallium(I) ($\chi_P = 1.62$), the same being true for lead (values quoted in Table 2.5 are for the higher element oxidation state where differences are significant). With regard to hybridization, it has long been recognised that, in the case of carbon, the order of electronegativity is $sp > sp^2 > sp^3$, the reason being that electrons are more tightly held (lower in energy) in s orbitals, and more advanced treatments of electronegativity assign specific electronegativities to particular orbitals or 'valence states' of atoms.

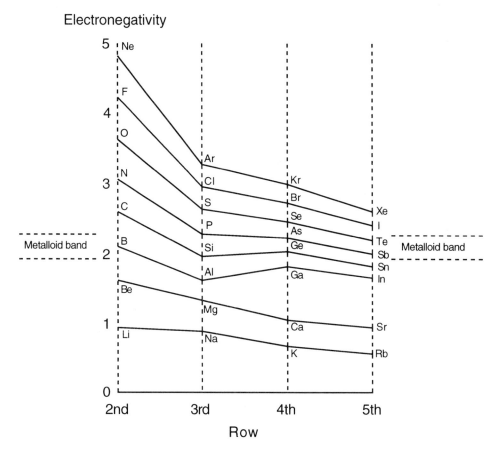

Fig. 2.6 Allen electronegativities as a function of group.

As a final point concerning electronegativity, we should note that there is clearly a relation to other features such effective nuclear charge, ionization energy and electron affinity and even atomic size, as mentioned earlier, but this should be expected since all are related properties; they are not independent of each other.

Finally, electronegativity is sometimes regarded as a chemical or electrical potential of an atom which is often a useful guide to reactivity; we will return to this matter in Chapter 6.

2.6 Orbital energies and promotion energies

Orbital energies and the energies of electron promotion from the ground state to a 'valence state' are important atomic properties and are useful in understanding certain aspects of periodicity. However, there is sometimes confusion between the two which can lead to ambiguous statements, particularly in discussions about the so-called inert pair effect which we will look at in more detail later.

Figure 2.7 shows the calculated one-electron energies of the valence s and p orbitals for the p-block elements. We can note first that within a particular group, the orbital energies increase (become less negative) as the group is descended (i.e. electrons become easier to remove) and also that, in general, the s–p energy separation decreases. The only exception to this generalization is with some of the elements of the 4p row. Thus, gallium and germanium, in particular, have larger s–p separations than aluminium and silicon respectively, with the energy of the s orbitals lower than their congeners in the 3p row. Moreover, for the remaining elements, As, Se, Br and Kr, the s orbital energy is lower than would be anticipated on the basis of the gradual trends expected from elements in other rows. This observation is another consequence of the increased effective nuclear charge of the 4p elements, discussed above, and it is clear that the effect on s orbitals is larger than that on p orbitals which is due to the greater penetration of s orbital electron density into the inner electron core. The parallel with trends in electronegativity is obvious; indeed the similarity with the Allen scale is a result of the fact that this scale is related to orbital energies.

One-electron orbital energies are used extensively in many types of molecular orbital calculations but an underlying assumption is that orbital energies do not change as a function of orbital occupation (Koopmans's theorem). This is a considerable over-simplification and care must be exercised in discussions involving these energies.

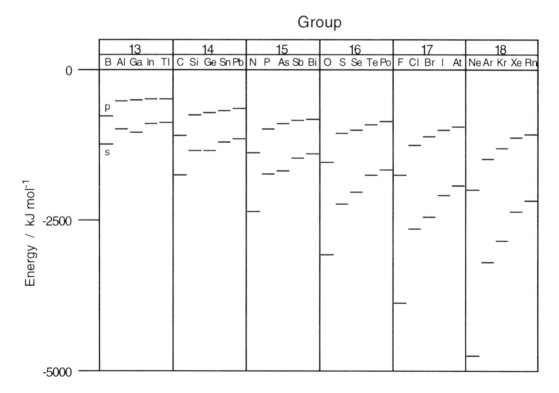

Fig. 2.7 Calculated one-electron energies of the valence s and p orbitals for the p-block elements. For each element, the s orbital is lower in energy than the p orbitals as illustrated for boron.

We might also expect a similar situation for the 6p elements as a result of the presence of the 4f row and relativistic effects. If the effect is present at all, it is clearly not as dramatic as for the 4p row but the s–p separations for thallium and lead are very similar to those of indium and tin respectively; for the later 6p elements it is not so obvious.

The other obvious trend is that both s and p energies decrease markedly as a period is crossed from left to right, particularly for the first row (2p) elements, and also that the s–p separation increases considerably. This reflects the very large increase in effective nuclear charge on moving from left to right, and the more pronounced effect on s orbitals is also apparent.

One of the problems with one-electron orbital energies is that they do not take account of so-called electron correlation effects, i.e. the effect that moving an electron from one orbital to another has on the energies of other electrons. This reorganization energy, as it is sometimes called, can be quite substantial, particularly for the heavier elements, and is revealed in the experimentally obtained electron promotion energies. Figure 2.8 shows the promotion energies for the group 13 and 14 elements, these being the energies required for the electronic transitions ns^2np^1 to ns^1np^2 and for ns^2np^2 to ns^1np^3 respectively. Perhaps the most notable and somewhat unexpected feature is that the promotion energies of the 6p elements thallium and lead are the highest in their groups indicating that the reorganization energy is considerable. We will return to a discussion of this matter in Chapter 4.

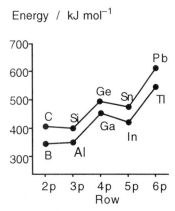

Energy / kJ mol^{-1}

Fig. 2.8 Experimental electron promotion energies for the elements of group 13 and 14. The actual values in Fig. 2.8 are (kJ mol^{-1}) B, 346; Al, 349; Ga, 456; In, 420; Tl, 543; C, 405; Si, 400; Ge, 504; Sn, 476; Pb, 609; taken from Dasent (1965).

It may seem strange to be talking about electron velocities and, by implication, of electrons as particles after all the discussion in Chapter 1 on electrons as waves. It is important to remember, however, that the concept of wave–particle duality implies that electrons are neither waves nor particles. In some situations, such as the electronic structure of atoms, the wave model is the most appropriate whereas in others a particle model is more useful.

2.7 Relativistic effects

One of the consequences of special relativity is that objects moving close to the speed of light increase in mass (relative to a stationary observer). The importance of this to chemistry is that for electrons in the heavier elements of the periodic table, velocities can be an appreciable fraction of the speed of light, as a result of the high nuclear charge, and these electrons are therefore subject to so-called *relativistic effects* (strictly speaking, electrons in all atoms are subject to relativistic effects but the magnitude of the effect is only appreciable for the heavier elements). Thus, as the velocity, and hence mass, of the electron increases, the effective size of the orbital decreases (direct relativistic orbital contraction) and its energy is lowered. The effect is most pronounced for electrons in s orbitals, since it is these orbitals which have the greatest density near the nucleus, rather less marked for p orbitals and relatively unimportant for d and f orbitals, although these latter two types of orbital are affected indirectly by the contraction of the s and p orbitals such that they are less well shielded and hence more diffuse than they would otherwise be (indirect relativistic orbital expansion). For the heavier p-block elements Tl through Rn there is an appreciable stabilization of the electrons in the 6s orbital which has important consequences for the chemistry of these elements.

3 Periodicity in the properties of the elements

In this chapter we will look at the properties of the elements themselves and attempt to understand the origin of the observed periodic trends. The first, and perhaps most important, point to note is that the p-block is the only part of the periodic table which contains non-metals; all of the s-, d- and f-block elements are metals. Moreover, a look at Table 3.1 reveals that within the p-block, the metals are found in the bottom left-hand corner whilst the so-called metalloids or semi-metals, shown in circles, lie along a diagonal from top left to bottom right with the non-metals, shown in bold type, occupying the top right of the Table. There is, therefore, a trend from metallic to non-metallic character as we move from bottom left to top right. This pattern correlates precisely with the trend seen for the element electronegativities discussed in the previous chapter and, as we shall see, it is electronegativity which we can use to rationalize the observations described above.

Hydrogen is something of a special case. It is certainly a non-metal under normal conditions, its electronegativity (2.20, Pauling; 2.300, Allen) being such as to place it in the non-metal class. It is therefore somewhat out of sorts with other s-block elements, but its unique chemistry is such that it is often placed in group on its own.

Table 3.1 The s- and p-block part of the periodic table illustrating the regions of metals, metalloids and non-metals; non-metals in bold, metalloids in circles

H								He
Li	Be		B	**C**	**N**	**O**	**F**	**Ne**
Na	Mg		Al	Si	**P**	**S**	**Cl**	**Ar**
K	Ca		Ga	Ge	As	**Se**	**Br**	**Kr**
Rb	Sr		In	Sn	Sb	Te	**I**	**Xe**
Cs	Ba		Tl	Pb	Bi	Po	**At**	**Rn**

The categories illustrated in Table 3.1 are not without some ambiguity particularly with regard to which elements are classed as metalloids; borders are often open to interpretation. Thus, bismuth and polonium are sometimes described as metalloids.

Before we look for an explanation of why particular elements have the properties they do, it is as well to be clear about what is meant by the terms 'metal, metalloid and non-metal'. For our purposes, it is probably sufficient to concentrate on the electrical properties whereby we can define metals as conductors, non-metals as insulators and the metalloids as semiconductors. These properties can be understood in terms of the so-called band theory of solids which we can illustrate with the aid of the simple diagrams shown in Fig. 3.1.

Band theory is essentially the molecular orbital theory of solids. For small molecules, we get a small number of discrete orbitals (bonding, antibonding etc.) equal to the number of atomic orbitals from which they are derived. For solids, the number of atomic orbitals is very large and so the resulting large number of molecular orbitals tend to be very closely spaced in energy and usually occur in groups which are called bands.

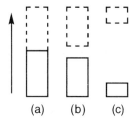

(a)   (b)   (c)

Fig. 3.1 Diagrams showing the band structure of (a) a conductor, (b) a semiconductor and (c) an insulator. Orbital or band energy increases in the direction of the arrow.

For the s- and p-block metals in particular, the number of valence electrons is generally less than the number of valence orbitals which necessarily results in the bands being only partially filled.

For conductors, Fig. 3.1(a), we have a partially filled band of orbitals (the bold lines indicate that part which is filled, the dashed lines the part which is unfilled) and electrons can readily be excited into unfilled levels in which they can move easily through the solid. An important characteristic is that the resulting high electrical conductivity decreases slightly with increasing temperature (due to increased scattering of electrons resulting from lattice vibrations). For semiconductors, Fig. 3.1(b), a filled band is separated from an unfilled band by a small energy gap, the so-called band-gap. Electrons are easily promoted by thermal excitation from the filled levels (in which they cannot move) to the unfilled, delocalized, so-called conduction band resulting in a dramatic increase in conductivity with temperature (this outweighs the tendency for conductivity to decrease with temperature as found in metals). For insulators, Fig. 3.1(c), the band-gap is too large to allow for easy promotion of electrons and so no conduction occurs.

Returning now to the elements, a comparison of Table 3.1 and the Allen electronegativities shown in Table 2.6 in the previous chapter, reveals that elements with an electronegativity smaller than about 1.9 are metals whereas those with electronegativities greater than about 2.2 are non-metals. Thus the metalloids all fall within the fairly narrow range of electronegativity of 1.9 to 2.2 and we can use this range of electronegativities to define the metalloid band which is shown in Fig. 2.6. Let us see why it is that electronegativity is important in this regard.

In the previous chapter we saw that the energy separation between the valence s and p orbitals is related to electronegativity such that those elements with the largest electronegativity also have the largest valence s–p energy separation. Thus, if we consider the elements with low electronegativity, the s and p orbitals are close together in energy and are therefore able to easily mix or overlap with each other (this is true whether we are considering s–p mixing on a particular atom, i.e. hybridization, or s–p overlap between atoms in the formation of molecular orbitals). The result of this ease of mixing is that when a number of atoms are placed in close proximity to each other, there is extensive orbital overlap between atoms (s–s, p–p and s–p) and closely spaced energy bands with a relatively large width are formed. This overlap is made easier by the fact that orbitals of the elements of low electronegativity tend to be fairly diffuse with correspondingly large radial extensions since the electrons are not tightly held. Wide and closely spaced bands tend to overlap with each other resulting in one large band which is only partly filled with electrons. This is a characteristic of metals wherein the bonding is highly delocalized, and the fact that they are only partly filled results, as we have seen above, in electrical conductivity.

For elements with a large electronegativity, the s–p energy separation is correspondingly large. Extensive s–p mixing and overlap between the orbitals of neighbouring atoms therefore tends to be reduced with the result that extensively delocalized bonding is disfavoured with respect to the formation of localized covalent bonds. For the elements nitrogen, oxygen and fluorine, this includes multiple bonding and covalently bonded diatomics result, whereas for elements such as phosphorus and sulfur, single bonds predominate which result in larger molecular or polymeric structures. In these elements there are large energy separations between filled and unfilled orbitals or bands even for polymeric structures, and as a consequence, they are insulators. The fact that high electronegativity results in valence electrons being tightly bound is also a reason for delocalised bonding being unfavourable.

The situation for the metalloids is intermediate between these two extremes. The band structure of the polymeric or macromolecular structures which these materials adopt resembles that of the non-metals in that bands are either filled or unfilled, but the band gap is small and electrons are therefore easily promoted to the conduction band resulting in electrical conductivity.

An alternative rationale for the observed metallic or non-metallic properties of the elements is associated with what is called a Peierls distortion. Briefly, a Peierls distortion (the solid-state analogue of a Jahn-Teller distortion) is expected to occur in any situation where there is a partially filled band resulting in the formation of a band-gap and an overall lowering of the energy of the filled orbitals at the expense of raising the energy of the unfilled ones. From a structural point of view, this results in a delocalised structure distorting to a structure of lower symmetry in which the bonding is more localised.

In principle, any structure with a partially filled band, i.e. a metal, should distort but the magnitude of the distortion is all important. This magnitude depends on the number of electrons present (the more you can stabilise, the better) but also, crucially, on the element electronegativity and its relation to valence orbital energies. Thus, if we start with a hypothetical delocalised structure, we can see what happens as we move from left to right in the periodic table. On the left hand side where electronegativity is small, all the orbitals are close in energy and there is little to be gained from lowering the energy of some of them and this factor, coupled with the small number of electrons present, means that there is little driving force for any distortion to a structure with more localised bonding and a metallic structure is prefered. Moreover, any small distortion that might be favoured energetically is likely to be overcome by thermal vibrations in the lattice. For elements on the right, however, the fact that the valence orbital energies are widely separated means that a lot can be gained by distorting a hypothetical delocalised structure to a more localised one since it is possible to lower some orbitals in energy by a considerable amount. This factor, coupled with the larger number of valence electrons present results in a large driving force for distortion to a structure with localised rather than delocalised covalent bonding, i.e. a non-metallic form.

It will now be useful to look at the structures of the elements in a little more detail in order to better appreciate some of the points discussed above although we will not deal with this topic comprehensively, especially the extensive allotropy found for some elements like phosphorus and sulfur.

A fuller discussion of Peierls and Jahn-Teller distortions is given in Albright, Burdett and Whangbo (1985).

The term allotrope is used here to distinguish between different forms of the same element, for example white phosphorus, P_4, and polymeric red phosphorus. Polymorphism is used to distinguish between different crystalline modifications of the same molecular form such as the orthorhombic and monoclinic forms of S_8.

3.1 Groups 1 and 2

With the exception of hydrogen, which exists under normal conditions as a gas containing diatomic molecules, all the elements of groups 1 and 2 are metals and have typical metallic structures; either body-centred cubic, face-centred cubic or hexagonal close-packed.

3.2 Group 13

The only non-metallic element in this group is boron which is generally classed as a metalloid. Elemental boron exists as a number of different allotropes/polymorphs each of which is a polymeric or macromolecular solid. Most of these forms contain B_{12} icosahedra, linked together in complicated three-dimensional arrangements. This structural complexity is a manifestation of the fact that boron is electron deficient in the sense that it has four valence orbitals but only three valence electrons. This type of electron deficiency, which inevitably occurs on the left-hand side of the periodic table, usually results in delocalized metallic bonding but, as noted above, boron is too electronegative to be a metal and instead forms more localized covalent bonds resulting in metalloid characteristics (note, however, that as a result of boron's electron deficiency, many of the B–B bonds are multicentre bonds rather than the more usual 2-centre-2-electron bonds found further to the right in the table).

The remaining elements in this group are all metals, aluminium and thallium having typical close-packed metallic structures with twelve nearest neighbours. Indium has a slightly distorted close-packed structure with four atoms marginally closer than the other eight but gallium has a rather unusual structure in which each gallium atom has one neighbour (2.44 Å) considerably nearer than the others which occur in pairs at 2.70, 2.73 and 2.79 Å. In fact the structure of solid gallium is quite similar to that of solid iodine, which consists of discrete I_2 molecules (see later), and it has been suggested that weakly bonded Ga_2 molecules are present in the solid phase of this element.

Group 13 is often considered to be the worst group in the periodic table for trends since, in as much as any are apparent, they often seem to be rather irregular. Note, however, that we can use electronegativity to rationalize some of what is outlined above. Boron is too electronegative to be a metal but the least electronegative elements, aluminium, indium and thallium, adopt typical metallic structures albeit slightly distorted in the case of indium. Gallium, the second most electronegative element in the group, has a structure in which incipient bond localization (in the form of the Ga_2 units mentioned earlier) typical of the metalloids is present.

3.3 Group 14

Group 14 is perhaps the group that best illustrates the trend from non-metallic to metallic properties as the group is descended reflecting the corresponding decrease in electronegativity. Carbon is a non-metal and is found in two common allotropic forms namely graphite and diamond, the structures of which are illustrated in Figs. 3.2 and 3.3. In diamond, each carbon atom is surrounded by four nearest neighbours at the vertices of a tetrahedron, the atoms being linked by strong, directional, 2-centre-2-electron covalent bonds. Graphite, in contrast, is a layered structure consisting of planar hexagonal sheets stacked on top of each other. Each carbon atom is surrounded by three nearest neighbours and extensive element–element π-bonding is important resulting from the overlap of one p orbital per carbon atom aligned perpendicular to the hexagonal sheet. It is this feature which gives rise to graphite being a conductor and, in this regard, makes it something of an exception as a non-metal.

Both diamond and graphite occur as two polymorphs. Cubic diamond is the most common (Fig. 3.2) in which all six-membered rings have a chair conformation whereas in hexagonal diamond, some rings adopt the slightly less stable boat form. The most stable form of graphite is the hexagonal form (Fig. 3.3) but a rhombohedral modification is also known, the two types differing in the stacking arrangement of the hexagonal sheets. Note that the terms cubic and hexagonal etc. refer to the symmetry of the crystal lattice and not to the coordination geometry.

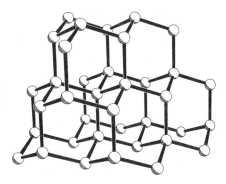

Fig. 3.2 Part of the structure of cubic diamond.

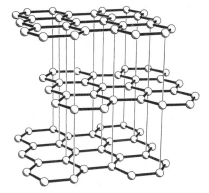

Fig. 3.3 Part of the structure of hexagonal graphite.

Extensive π-bonding is also a feature in the structures of the newly discovered molecular forms of carbon, the fullerenes, of which the archetype is Buckminsterfullerene or C_{60} shown in Fig. 3.4.

Silicon and germanium both have the cubic diamond structure and are classified as metalloids largely as a result of their semiconducting properties. These are elements of intermediate electronegativity in which localized covalent bonding occurs but for which the valence s–p energy separations are small enough to result in fairly closely spaced bands. We might note that diamond, which has a similar band structure to Si and Ge, is an insulator due to its much larger band gap resulting from the greater s–p energy separation although it is often referred to as a high band-gap semiconductor (rather than a non-metal) in the field of solid state physics; it all depends on your point of view.

The much larger band-gap in diamond, as opposed to that in silicon or germanium, is also a result of the smaller size of carbon; the C–C bonds are therefore significantly shorter than the corresponding Si–Si and Ge–Ge bonds which leads to correspondingly better orbital overlap.

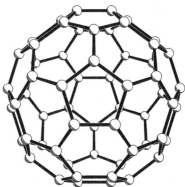

Fig. 3.4 The structure of Buckminsterfullerene, C_{60}.

Another difference between diamond and silicon or germanium is that pure diamond is clear and colourless (colours in natural diamonds are due to impurities) whereas silicon and germanium are black. These properties also result from the differences in the band-gaps; visible light has sufficient energy to promote electrons from the lower to the higher band in silicon or germanium (but not in diamond) and hence is strongly absorbed.

Tin lies close to the metalloid–metal border and exists as two allotropes, grey and white tin, the former having the cubic diamond structure. The latter has a much less regular structure, still with four nearest neighbours (slightly further away than in grey tin), but with additional tin atoms much closer than the second nearest neighbours in the grey tin structure making white tin more dense and more metallic in its properties. There is a suggestion that these two structures contain tin in two different oxidation states, +4 and +2 respectively, since grey tin dissolves in conc. HCl to give tin(IV) chloride, $SnCl_4(aq)$, whereas white tin dissolves to give tin(II) chloride, $SnCl_2(aq)$. This may seem a curious statement since we normally regard elements as having a zero oxidation state, but what is implied is that whereas in grey tin all four valence electrons are used in bonding, in white tin, only two are used (the metal is thought of here as an ionic structure with metal cations and electrons as the anions). This is an example of the *inert pair effect*, which we will discuss in detail in the next chapter, in which the valence s pair is not used in bonding.

The idea of an inert s pair is also relevant to the structure of lead, which although that of a typical close-packed metal, has particularly long Pb–Pb distances indicative of the presence of a non-bonding 6s pair of electrons. The interatomic separations in the structures of In and Tl are also long, indicating that a similar situation occurs here as well. In general, we expect longer distances when the s pair is not involved in bonding; for example, the ionic radius of Sn(II) is larger than that of Sn(IV), see later.

3.4 Group 15

Nitrogen is a true non-metal and exists exclusively as a diatomic molecule, N_2, in which the two nitrogen atoms are held together by an extremely strong triple bond. This strong, directional, localized covalent bonding is to be expected for such an electronegative element, and we can also note the presence and importance of N–N π-bonding.

In contrast to nitrogen, phosphorus does not exist as P_2 molecules except under conditions of high temperature and low pressure in the gas phase. The stable elemental forms are white (or yellow) phosphorus, which contains tetrahedral P_4 molecules, and a variety of polymeric allotropes such as red, black and violet phosphorus. All forms of phosphorus contain three-coordinate phosphorus atoms in which each phosphorus is linked to three others by single P–P bonds. This feature typifies an important difference between the chemistry of phosphorus and nitrogen, and between the first and second row elements in general, in that there is a reluctance of the heavier elements to form stable element-element π bonds resulting in a prevalence of element–element single bonding or *catenation*, a topic which we will discuss in more detail later. The structure of orthorhombic black phosphorus, the most thermodynamically stable form, is illustrated in Fig. 3.5(c) and consists of layers of linked six-membered rings in the chair conformation, whilst that of a closely related higher pressure rhombohedral form is shown in Fig. 3.5(d). At even higher pressures, a cubic structure analogous to α-polonium is adopted (see later). Red phosphorus is amorphous and violet phosphorus, although crystalline, has a rather complicated structure.

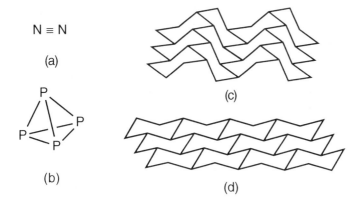

Fig. 3.5 (a) Dinitrogen, N_2, (b) tetraphosphorus, P_4, (c) a section of orthorhombic black phosphorus, and (d) a section of a layer in rhombohedral black phosphorus.

The common α forms of arsenic, antimony and bismuth are isostructural and consist of hexagonal sheets similar to that shown for rhombohedral black phosphorus, see Fig. 3.5(d). Other allotropic forms also exist, two of particular interest being high pressure metallic forms which for antimony is a hexagonal close-packed structure and for bismuth is ζ-Bi in which each bismuth is surrounded by eight nearest neighbours in a body-centred cubic lattice.

Isostructural means that the structures are the same.

Before leaving the group 15 elements we should note an interesting trend in the interatomic distances in the isostructural rhombohedral black phosphorus, α-As, α-Sb and α-Bi. As mentioned above and shown in Fig. 3.5, each phosphorus atom in rhombohedral black phosphorus has three nearest neighbours arranged at the vertices of a trigonal pyramid. In addition, there are three other atoms at a greater distance *trans* to the three primary bonds (in the next sheet) such that the overall coordination is octahedral with three short and three long bonds. The observed trend is that as the group is descended, the ratio of the long to short bonds decreases such that the bonds become more nearly equal in length. This is a definite trend from localized covalent bonding in a sheet structure towards a more delocalized, three-dimensional metallic bonding and is consistent with the decreasing electronegativity of the elements as the group is descended.

3.5 Group 16

The most stable form of oxygen is dioxygen, O_2, which is a gas under ambient conditions and in which the two oxygen atoms are held together by a strong multiple bond. The only other form of this element is ozone, O_3, (which is also a gas) with a bent, triatomic structure in which O–O multiple bonding is also important.

Polymorphism is also common in the elemental forms of sulfur. Thus S_8 exists in one orthorhombic and two monoclinic modifications.

$$O = O$$

(a)

(b)

(c)

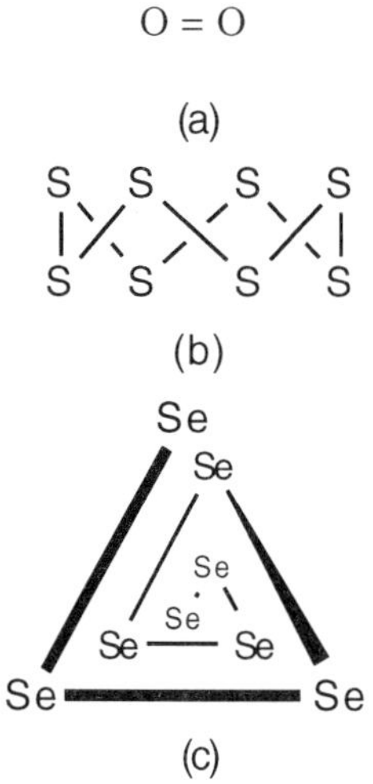

Fig. 3.6 (a) Dioxygen, O_2, (b) cyclo-octasulfur, S_8, and (c) part of a chain of helical selenium, Se_x.

In contrast, sulfur does not exist as S_2 or S_3, except at high temperatures, but rather as S–S singly bonded species, the number of allotropes of which is extremely large. We will not dwell on the details here except to say that besides the common form of sulfur which consists of S_8 rings (Fig. 3.6), many other rings of different sizes have been characterized together with polymeric, fibrous forms which contain helical chains of sulfur atoms. Both oxygen and sulfur are non-metallic as expected from their electronegativity.

Selenium also has properties consistent with its classification as a non-metal and its allotropy, although much less extensive than that of sulfur, comprises modifications which contain Se_8 rings and a hexagonal form in which the selenium atoms form unbranched, helical chains (Fig. 3.6). A structurally very complex vitreous form of selenium is also known in which rings of up to 1000 atoms are thought to be present. Tellurium, in contrast, exists as only a single allotrope which is isostructural with hexagonal selenium. A similar trend in bond distances to that noted for the heavier group 15 elements is also apparent for the hexagonal forms of selenium and tellurium. Thus the ratio of the short, covalent bond lengths to the longer interchain distances is smaller for tellurium than for selenium, again indicating a trend to metallic character.

The final element in this group is polonium which is the least electronegative and most metallic in character. It has a structure in which each atom is surrounded by six nearest neighbours with a simple cubic unit cell, the only element in the periodic table for which this is found.

3.6 Group 17

Isomorphous means that the crystal structures are the same.

The halogens all exist as diatomic molecules under normal conditions, F_2 and Cl_2 being gases, Br_2 a liquid and I_2 a solid. The low temperature solid state structures of Cl_2 and Br_2 are isomorphous with I_2 and all are layer structures as illustrated for I_2 in Fig. 3.7. The solid-state structure of F_2, although different, is also layered (the structure of astatine is not known as a result of its high radioactivity and consequent handling problems).

A number of further points are worth mentioning with regard to the solid-state structures of the halogens. Firstly, the ratio of the short, covalent bonds to the longer, intermolecular distances decreases as the group is descended in keeping with what we have seen in some of the preceding groups. This is illustrated in Table 3.2 below (distances in Å, the value for F_2 is in parentheses since it has a different crystal structure).

Table 3.2 Bond lengths (Å) for the halogens in the solid-state

X	X–X	X⋯⋯X Intralayer	X⋯⋯X Interlayer	Smallest ratio X⋯X/X–X
F	1.49	3.24	2.84	(1.91)
Cl	1.98	3.32, 3.82	3.74	1.68
Br	2.27	3.31, 3.79	3.99	1.46
I	2.72	3.50, 3.97	4.27	1.29

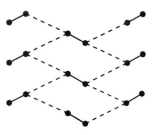

Fig. 3.7 A section of a layer in the solid-state structure of diiodine, I_2.

We should also recognise that van der Waals or induced dipole-induced dipole interactions increase as the elements become more polarizable on descending the group.

This observation is consistent with a trend towards increasing metallic character as a result of decreasing element electronegativity and in this regard it is interesting to note that under high pressures, solid iodine does become metallic as evidenced by the onset of electrical conductivity. We can think of this as forcing the atoms together to such an extent that orbital overlap becomes sufficient for extensive delocalization to occur which results in metallic bonding. A similar situation is thought to occur for dihydrogen, H_2, under very high pressures as well as for a number of metalloids.

Another point to note is that the I–I bond distance in solid iodine (2.72 Å) is considerably longer than that in gaseous iodine (2.67 Å). The longer bond in the solid state is undoubtedly a result of the closeness of the intermolecular contacts, a feature we shall meet again when dealing with the compounds of iodine (and other elements) and discuss at that time.

It is also noteworthy that the halogens become more intensely coloured as the group is descended; F_2 is essentially colourless, Cl_2 is pale green, Br_2 is deep red-brown and I_2 is dark purple which we shall see is also a feature of the compounds formed by these elements, and that an explanation has much in common with that for the trends in bond distances described in the previous paragraph.

3.7 Group 18

The group 18 elements are all monatomic gases under ambient conditions and are entirely non-metallic. Whilst this certainly reflects their high electronegativity, the general lack of reactivity resulting from a filled valence shell is also an important factor.

In the solid-state the group 18 elements are close-packed.

3.8 General features

An interesting insight into the relationships between many of the apparently very disparate structures of the elements in the p-block can be had by considering the numbers of valence electrons present. For example, Figure 3.8(a) shows a section of the structure of cubic diamond (which is also the structure found for silicon, germanium and grey tin). Each carbon atom forms four 2-centre-2-electron C–C bonds and there are no lone pairs of electrons. If we now move one group to the right (i.e. to group 15) we have added one valence electron per atom or two electrons per pair of atoms. This leads to the breaking of one bond (out of four) for each atom such that this bond pair becomes two lone pairs, one on each atom. The resulting atomic coordination is trigonal pyramidal (three bond pairs and one lone pair). If the bonds broken are the vertical bonds in Fig. 3.8(a) (of which there is one per atom), the resulting structure comprises parallel two-dimensional sheets of fused six-membered rings in the chair conformation as shown in Fig. 3.8(b). This is the structure of rhombohedral black phosphorus, α-As, α-Sb and α-Bi.

It should be noted that the relationship of the layers in rhombohedral black phosphorus to each other is not exactly the same as the corresponding layers in cubic diamond. In rhombohedral black phosphorus the sheets have moved relative to each other to avoid close contacts between the lone pairs. Similar dislocations or translations occur in the tellurium and diiodine structures with respect to the positions shown in Fig. 3.8 in order to minimise lone pair-lone pair contacts.

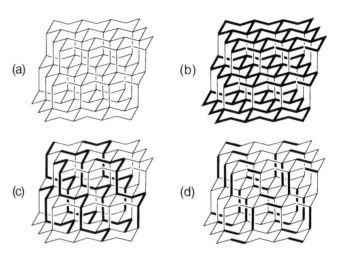

Fig. 3.8 Four diagrams showing the structure of cubic diamond. (a) represents diamond and other group 14 structures while (b), (c) and (d) represent how the structures of the group 15, 16 and 17 elements can be derived from the diamond structure.

If we move to group 16, there are now another two electrons per pair of atoms which again leads to the breaking of one bond per atom resulting in two-coordination (two bond pairs and two lone pairs). This can give rise to a helical structure as found in elemental tellurium (and one allotrope of sulfur and selenium) and is shown in Fig. 3.8(c).

Finally, the addition of a further electron per atom takes us to group 17 and we break another bond so that each atom is now one-coordinate with three lone pairs. This is the structure of dichlorine, dibromine and diiodine in the solid-state and is represented in Fig. 3.8(d).

For the elemental structures described here the number of bonds formed by any particular element is given by the so-called 8–N rule where N is the number of valence electrons.

An alternative approach to rationalising the relationships between the structures of the p-block elements is in terms of Peierls distortions mentioned earlier. Thus if we take a simple cubic lattice as shown in Fig. 3.9(a) and which represents the structure of α-polonium, a metal, we can derive the structures of helical selenium and tellurium by removing four lines per lattice point (Fig. 3.9(b)) together with those of orthorhombic black phosphorus and rhombohedral black phosphorus plus α-As, Sb and Bi by removing three lines per lattice point in the ways shown in Figs. 3.9(c) and (d) respectively. A full treatment of this approach is beyond the scope of this book, but it is possible to derive the structures of most of the elemental forms of the p-block elements along these lines (boron is a bit of a problem).

Figs. 3.9(c) and (d), in particular, illustrate how closely related are these two forms of phosphorus.

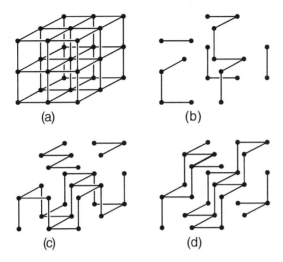

(a)   (b)

(c)   (d)

Fig. 3.9 Four diagrams showing (a) a simple cubic lattice and how the structures of (b) tellurium, (c) orthorhombic black phosphorus and (d) rhombohedral black phosphorus can be derived from this lattice.

3.9 Binding energies

A final point of interest whilst dealing with the elements themselves concerns the so-called binding energies (or atomization energy) which relate to the magnitude of the interatomic forces which bind the elements together.

For example, in a diatomic molecule, the binding energy is defined according to Eqn 3.1 whereas for a macromolecular solid, Eqn 3.2 is appropriate.

$$X_2 (s) \;\rightarrow\; 2X (g) \qquad (3.1)$$

$$X_n (s) \;\rightarrow\; nX (g) \qquad (3.2)$$

A graphical representation of the binding energies for the elements of the first two rows is shown in Fig. 3.10 from which it is apparent that the energies peak in group 14 and fall to zero for the group 18 elements. The nature of this graph simply reflects the number of valence electrons available for bonding. The trend from lithium to carbon in the first row is a consequence of the number of valence electrons increasing from one to four. For nitrogen, although there are five valence electrons, only three are used in bonding (the remaining two are a lone pair) and so the binding energy is similar to that of boron (note that this is true even though elemental boron and nitrogen have quite different structures and very different melting and boiling points). Similarly, sulfur with six valence electrons uses only two in element–element bonding and therefore has a similar binding energy to beryllium. In general, the reactivity of an element will increase as its binding energy decreases although the noble gases in group 18 are an obvious exception.

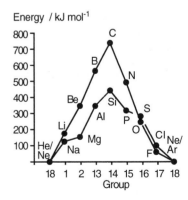

Fig. 3.10 The binding energies for the elements of the first two rows.

4 General features of s- and p-block element compounds

Before considering the various classes of compounds, which we will do in the next chapter, it will be useful to look at some general aspects, such as trends in common oxidation states, and the effect of atomic size on coordination numbers (Sections 4.1 and 4.2) which, together with properties such as electronegativity described in Chapter 2, will form the basis of our understanding of s- and p-block element structure and reactivity. In addition we will look at trends in bond strengths (Section 4.3) and finally at some general observations regarding compound types using the so-called van Arkel-Ketelaar diagram (Section 4.4).

4.1 Oxidation states, valence and the inert pair effect

As a first step, we should briefly review the method by which numbers for the oxidation state of an element are derived.

If we consider CF_4, then it is usual in assigning oxidation states to partition the elements according to their electronegativity such that we obtain C^{4+} and $4F^-$. This is, therefore, a compound of carbon(IV) or carbon in the +4 oxidation state, although it is important to stress that this is only a formalism and does not imply that CF_4 is in any way an ionic compound: fluorine is present as F(–I) or fluoride. Similarly, NF_3 is a compound of nitrogen(III) since fluorine is more electronegative than nitrogen. In contrast, for NH_3, nitrogen is more electronegative than hydrogen and so the atoms would be partitioned as N^{3-} and $3H^+$ thereby giving nitrogen(–III).

If we now consider N_2F_4, then removing four fluorines as F^- leaves N_2^{4+} or $2N^{2+}$ and so dinitrogen tetrafluoride formally contains nitrogen(II). We should note, however, that in all of the above three nitrogen compounds, the nitrogen atoms are singly bonded to three other elements and may therefore be classed as *trivalent*. This distinction between formal oxidation state and valence is an important one which we will meet again later. Similarly, the compound N_2F_2 formally contains nitrogen(I) centres (N_2^{2+}) but the nitrogen atoms are still trivalent forming a single bond to fluorine and a double bond with each other.

Let us now turn to some general features regarding the common oxidation states which are found for the s- and p-block elements and why some states are much more common than others. Probably the most notable factor is that for elements with more than one available oxidation state, the formal values almost always differ by two units as indicated in Table 4.1.

The values in Table 4.1 do not constitute an exhaustive list and there are some important exceptions such as NO and NO_2, which are formally

There is a sense, using the examples NF_3, N_2F_4 and N_2F_2, in which the oxidation state formalism seems unwieldy and indeed some (notably Green, see *J.Orgamomet.Chem.*, (1995), **500**, 127) have argued that this is so. Alternative constructions which concentrate on the electron count around the atom in question have been developed, although they have so far mostly been used in the context of d-block chemistry. Nevertheless, while aware of any possible limitations of the oxidation state formalism, we will use it here to illustrate certain aspects of periodicity.

In solid-state chemistry, a slightly different concept of so-called bond-valence is often used since the assignment of valence and oxidation state is not always straight-forward. Thus, the valence V_A of an atom A in a compound AX_n is distributed over all the A–X bonds according to the valence sum rule $\Sigma s_{AX} = V_A$ where s_{AX} is the bond valence of a particular bond and is correlated with the length of that bond according to $s_{AX} = \exp[(r_o - D_{AX})/b]$ where r_o and b are empirically detemined parameters, and V_A is equal to the formal oxidation state of A.

nitrogen(II) and (IV) respectively, and these we should consider at this point. The key aspect to note is that elements in their common oxidation states listed in Table 4.1 contain no unpaired electrons whereas the three compounds just mentioned are all radicals, i.e. they *do* contain unpaired electrons. The reason that radicals are not all that common in s- and p-block chemistry is that the unpaired electrons would generally reside in p or sp^n type hybrid orbitals which have fairly good directional properties. Overlap between these orbitals therefore tends to be favorable and reasonably strong bonds are formed such that there is a strong tendency for radical species to dimerize unless the unpaired electrons are fairly extensively delocalized or the resulting bond is particularly weak for some reason.

Table 4.1 Common oxidation states for the s- and p-block elements

Group	1	2	13	14	15	16	17	18
Oxidation	1	2	3, 1	4, 2	5, 3,	6, 4,	7, 5,	6, 4,
state					-3	2,-2	3, 1,-1	2, 0

Delocalization of the unpaired electron is the situation in NO, where the unpaired electron resides in a N–O π^* orbital, but in NF_2, for example, the unpaired elecron is in a sp^2 hybrid and N–N bond formation occurs. Thus, the resulting N_2F_4, although formally a compound of nitrogen(II), contains trivalent nitrogen, as described above, and is diamagnetic, whilst nitric oxide, which is also nitrogen(II), contains genuinely divalent nitrogen and is paramagnetic; in fact N_2F_4 dissociates fairly readily in the gas phase to give NF_2 radicals and is not such a good example to use, the two fluorines lead to effective delocalization of the unpaired electron and shrinkage of the orbital containing that electron, but in general, R_2E species where E is a group 15 element almost always exist as R_4E_2. In the case of NO_2, the weak N–N bond in the dimer N_2O_4 is the reason why appreciable dissociation into NO_2 radicals occurs.

We should note that the rarity of radicals in s- and p-block chemistry is in stark contrast to the situation found in compounds of the d- and f-block elements. Here radicals are frequently encountered and common oxidation states often differ by only one unit, e.g. Fe(II) and Fe(III). This can be traced to the fact that the valence electrons in d- and f-block complexes are generally in d and f orbitals respectively, which are more diffuse and less directional than corresponding s–p hybrids. The tendency for compounds with unpaired electrons to dimerize with the formation of element–element bonds is therefore not so pronounced for this reason, although several other factors are also important such as the often relatively small radial extension of d and f orbitals.

Let us now turn to the reasons behind the occurence of the common oxidation state values for the s- and p-block elements. As we have just seen, part of the reason has to do with the formation of compounds without unpaired electrons. The most important factor, however, concerns the attainment of a closed-shell electronic configuration, by which is meant an electron configuration that resembles that of the nearest noble gas. This idea

Part of the reason for the reluctance of NO to dimerize to give N_2O_2, which does exist in the solid state, may also be attributed to the net change of zero in bond order occuring on dimerization. Thus the bond order in NO is 2.5 since the unpaired electron resides in a π^* orbital, i.e. exactly half that of N_2O_2 which is five in total. There is, therefore, no net increase in bonding upon dimerization, at least to a first approximation, as would more usually be the case in the dimerization of radicals.

was first enunciated by Lewis as the octet rule and its importance in rationalizing a large body of knowledge in s- and p-block chemistry should not be understated. Moreover, we shall see later that most apparent violations of the octet rule are, in reality, no such thing.

In groups 1 and 2, the oxidation states are 1 and 2 respectively resulting from loss of the valence s electrons and any higher oxidation states are precluded by the prohibitively large energies required to remove electrons from the next quantum shell down. The reason why compounds of the group 2 elements in the +1 oxidation state are not observed is less obvious, but thermodynamic calculations indicate that disproportionation reactions (Eqn. 4.1) for simple salts EX (X = halide, for example) are strongly exothermic being driven by the large lattice energy of the resulting EX_2 salt.

$$2\,EX \quad \rightarrow \quad E + EX_2 \quad (4.1)$$

If we consider group 13, the common oxidation states are +3 and +1. The former arises by formal loss of the three valence electrons, $ns^2\,np^1$, which gives a +3 centre with a noble gas core (remember that we are dealing with a formalism here; in TlF_3 it is reasonable to view this compound as predominantly ionic with an Tl^{3+} cation and three fluoride anions, but for the much more covalent InI_3 the charges are purely formal).

The +3 state is observed for all elements in this group, but the +1 state, which results from removal of the single valence p electron, is common only for the heavier elements, particularly so for thallium. In fact, the +1 oxidation state is probably more important than the +3 state in the chemistry of thallium, and we shall see that oxidation states of two less than the group maximum are commonly observed for the 6p elements. This is usually refered to as *the inert pair effect* and we should consider the origin of the effect at this point using thallium as an example.

For the 6p elements thallium, lead and bismuth, the point about lower oxidation states is certainly true; for polonium, astatine and radon, experimental data are rather scarce due to their radioactivity and consequent handling problems.

A standard and often quoted explanation is that the relatively large effective nuclear charge of thallium, resulting from the presence of the 4f and 5d elements, coupled with relativistic effects, leads to a substantial stabilization of the 6s pair relative to the 6p electron, and it is this factor which gives rise to the term 'inert pair' itself. However, this is only part of the explanation and, indeed, the term must be seen as something of a misnomer in this regard. Although the 6s pair is stabilized by these effects, it can hardly be considered to be too low in energy (i.e. core like) in view of the relevant ionization energies which are comparable with those of the lighter elements. This is illustrated in Table 4.2 in which values for the sum of the first three ionization energies (Σ IE 1–3) for the group 13 elements are presented.

Note that the trend in ionization energies in Table 4.2 mirrors that found for the electronegativities of these elements, although this is not surprising since the Mulliken electronegativity scale, for example, actually uses ionization energies to calculate electronegativities. Thus, the electronegativity of gallium is higher than aluminium and that of thallium is higher than indium.

Table 4.2 Values for the sums of the first three ionization energies for the group 13 elements (kJ mol^{-1})

Element	B	Al	Ga	In	Tl
Σ IE 1–3	6888	5140	5521	5084	5438

All the values in Table 4.2 are of a comparable magnitude and we therefore cannot argue that the 6s pair of thallium is unavailable for bonding because it is too low in energy. If this argument were appropriate at all, we would expect that, of all the group 13 elements, boron would exhibit an inert pair effect, since the s pair ionization energies are significantly larger for this element, a factor which is also apparent from the graph of orbital energies shown in Fig. 2.7, and one which would be particularly important for the elements further to the right in the 2p row.

A consideration of the experimental electron promotion energies required to achieve the trivalent state, shown in Fig. 2.8, is, perhaps, more appropriate, especially when dealing with covalent compounds, and here we should note that the value for thallium is the largest in the group. However, whilst this will clearly be important, it is not that much greater than the values for the other elements and cannot, therefore, constitute a complete explanation.

In fact, the real origin of the inert pair effect for thallium is probably a combination of two factors. The high promotion energy involved in achieving the valence state is one factor, but an equally important point is the weak bonds formed by this element as a result of its large size (see later). Thus for boron, the energy required to involve the 2s pair in bonding (the promotion energy) is more than offset by the formation of three strong covalent bonds and boron is therefore more likely to occur with an oxidation state of +3 rather than +1 or, more precisely, it will tend to be trivalent rather than monovalent. In the case of thallium, the opposite tends to be true and the formation of two extra, much weaker bonds commensurate with the +3 oxidation state, or trivalency, is not enough to compensate for the considerable promotion energy.

> The explanation for the high promotion energy of thallium is due to spin-orbit coupling effects rather than any difference in the orbital energies. We will not concentrate on the details here except to mention that spin-orbit coupling increases with *n* (the principal quantum number) and Z^*, this latter factor accounting for the apparent irregularity of the values in Fig. 2.8. See Dasent (1965).

We should also consider the significant difference in the radial extension or size of the 6s and the 6p orbitals as mentioned in Chapter 1. The importance in this context is that sp hybridization will be less efficient leading to weaker covalent bonding in situations where s orbital participation is required (such as trivalent thallium). We shall meet this idea again in later sections.

> A discussion of hybrid orbitals is given in Winter (1993).

In group 14, the common oxidation states are +4 and +2, (more generally, tetravalency and divalency) with the latter being most common for tin and lead. The inert pair effect is important for Pb(II), as it was for Tl(I), for the same reasons, and this oxidation state is the dominant one in lead chemistry, although we must be a little careful of generalizations of this type. For example, it is found that whilst lead(II) chloride, $PbCl_2$, is indeed more stable than lead(IV) chloride, $PbCl_4$, in line with the trends described above (see also Chapter 5), the opposite is true for the organometallic alkyl compounds; e.g. tetraethyl lead, $PbEt_4$, is stable whereas diethyl lead, $PbEt_2$, is not (a similar situation is found for the chlorides and alkyls of thallium(III) vs. thallium(I)). We will return to this matter in Section 4.3.

> We can note at this point that the inert pair effect should favour the zero oxidation state for mercury, the element one to the left of thallium in the periodic table. This accounts, at least in part, for why mercury metal is rather unreactive and why the element is a liquid at room temperature since it does not form strong element–element bonds.

In group 15, the common oxidation states are +5, +3 and −3 or more simply, compounds are generally pentavalent or trivalent. The inert pair effect can be (and often is) said to have some influence on the importance of bismuth(III) over bismuth(V), but with oxidation states as high as +5, there are other factors which are important in determining the relative stability of possible oxidation states. Thus, high positive oxidation states tend to be good oxidising agents since the high formal charge centres are good at attracting electrons and thereby being reduced. As such, high oxidation state centres are generally only found in association with fluorine or oxygen since these very electronegative elements are themselves difficult to oxidise. This feature of high element oxidation states being found in association with fluorine and oxygen is not just restricted to the p-block, it is common in the d- and f-blocks as well; the low oxidation states of the s-block elements require no special stabilising conditions. As an example, PF_5 is a known compound whereas PI_5 is not. While there is undoubtedly much truth in this view, it seems somewhat at odds with the discussion presented above for the stability of thallium(III) and lead(IV) alkyls vs. halides. A resolution of this matter involves a number of factors which will be addressed in Section 4.3.

We must be a little careful in considering the differences between PF_5 and PI_5. Thus whilst it is probably correct to say that PI_5 does not exist because P(V) would oxidize iodide, we must take size effects into consideration as well; iodide is large and it is probably not possible to place five iodines around a small P(V) centre.

In group 16 we find that the +6 oxidation state is, not surprisingly, found most often in conjunction with fluorine and oxygen although, as in the case of lead(IV), this is a generalization which can be taken a little too far. For example, tellurium hexamethyl, $TeMe_6$, has recently been synthesized and found to be relatively stable. The other stable states are the +4 and the +2, although in the case of oxygen, oxidation states greater than +2 are not observed. This is a consequence of the high electronegativity of oxygen and, strictly speaking, it will only be found in a positive oxidation state with elements of greater electronegativity which means fluorine; OF_2 is an example. In H_2O, oxygen is more electronegative than hydrogen and is therefore formally oxygen(–II). In both OF_2 and H_2O the oxygen is, of course, divalent and this is characteristic of the element. Sulphur is also often found as sulfur(–II) although in the presence of oxygen or fluorine, compounds of S(II), S(IV) and S(VI) are well established.

In group 17, the elements exhibit the widest range of oxidation states in the p-block. Fluorine, being the most electronegative element in the periodic table, is never found in formal postive oxidation states and exists as F(–I) in all of its compounds. The heavier halogens are found in oxidations states –1, +1, +3, +5 and +7 although the higher oxidation states are usually powerful oxidizing agents and are only found in conjunction with very electronegative atoms or groups such as fluorine.

In group 18 the most common oxidation state is zero, i.e. the atomic elements themselves, and this is the only state found for the elements He, Ne and Ar. Krypton is known in the +2 oxidation state as KrF_2, but only for xenon is the chemistry extensive with oxidation states of +2, +4 and +6 being observed (+8 is also known) albeit only in the presence of electronegative elements or groups. The reason for positive oxidation states being more common for the heavier elements simply reflects their lower ionization energies. The potentially extensive chemistry of radon being limited by the handling problems associated with its radioactivity.

Having considered the various groups in the s- and p-block let us turn to some more general points. One of these is the reluctance of the 4p elements to exhibit their highest possible or group oxidation state (c.f. the 6p elements mentioned earlier). This can be considered as another example of an inert pair effect, although it is not usually referred to as such, and, while it is not so apparent for the respective +3 and +4 states of gallium and germanium, arsenic(V), selenium(VI) and bromine(VII) are much less common oxidation states than phosphorus and antimony(V), sulfur and tellurium(VI) and chlorine and iodine(VII) respectively. Moreover, compounds of the 4p elements exhibiting the group maximum oxidation state are more powerful oxidizing agents than their lighter and heavier congeners.

The reasons for the high oxidation states of the 4p elements being uncommon are probably similar to those encountered for the related situation with the 6p elements. Thus, the increased effective nuclear charge associated with the presence of the preceding 3d row will make it more difficult to involve all the valence electrons in bonding and any extra bonds formed may not be sufficiently strong to compensate. With gallium and germanium, the effect is not so dramatic in that trivalent and tetravalent compounds respectively are not unusual since the bond energies are large enough to compensate for the increased promotion or ionization energies involved. For the later elements, however, the corresponding energies may well be too large to allow for sufficient compensation from the formation of extra bonds, but this is not a wholly satisfactory explanation for a number of reasons. From the point of view of orbital energies, the s–p separation is not excessively large (see Fig. 2.7) nor are the ionization energies for the s pair of electrons as shown in Table 4.3 (indeed, any irregularity is much more noticeable for the group 13 and 14 elements). With regard to promotion energies, these will involve s to d or p to d excitations and the energies are considerable, but it is not clear to what extent such excited states are chemically reasonable; in other words, the extent to which d orbitals are really involved. We will return to this matter of d orbital participation later.

It has been calculated that once an electron has been promoted into a d orbital, the electronegativity of the valence s and p orbitals increases dramatically since the d orbitals shield the s and p orbitals only poorly. Thus if d orbitals are involved, any such promotion would make it that much harder to involve all the s and p orbitals particularly in cases, such as the 4p row, where electronegativities are already high.

Table 4.3 Ionization energies (kJ mol^{-1}) for the s^2 electron pair for the 3p, 4p and 5p elements

Group 13	Al	Ga	In
Σ IE 2 + 3	4560	4940	4520
Group 14	Si	Ge	Sn
Σ IE 3 + 4	7580	7710	6870
Group 15	P	As	Sb
Σ IE 4 + 5	1 1230	1 0880	9660
Group 16	S	Se	Te
Σ IE 5 + 6	1 5510	1 4470	1 2490
Group 17	Cl	Br	I
Σ IE 6 + 7	2 0380	1 8490	1 6100

Another point to consider is the difference between the trends in oxidation states found in the p-block and those observed in the d- and f-blocks (there are no real trends to worry about in the s-block). As we have seen, there is a general trend to lower oxidation states as the groups in the p-block are descended (most notable in groups 13–15) which can be traced to the inert pair effect, a prominent aspect of which is the weaker bonds formed by the larger elements. In the d- and f-blocks, the reverse tends to be true, at least for the elements on the left-hand side of these blocks. Thus, third row transition metals and actinides often exhibit higher oxidation states than first row transition metals and the lanthanides respectively (e.g. Os(VIII) and U(VI) have no counterpart in the chemistry of iron and neodymium respectively). This can be traced to the difference in d and f compared to s and p orbitals and relativistic effects.

We can consider the situation in the p-block to be 'normal', explanations for which having been offered above. For the heavier elements of the d and f blocks, however, relativistic effects cause a contraction of the valence (and core) s orbitals, and, to some extent, the p orbitals which leaves the d and f orbitals exposed with the electrons in these orbitals feeling a reduced effective nuclear charge or smaller effective core potential. They are therefore easier to remove or ionize and so higher oxidation states are found in the chemistry of these elements. It is also the reason why covalent bond strengths tend to increase down a group in the d-block whereas the reverse is true in the p-block; we will return to this point later. The fact that higher oxidation states are less common as the d- and f-blocks are traversed from left to right simply reflects the increasing effective nuclear charge which means the electrons are, in general, harder to remove.

As mentioned previously, the trend to lower oxidation states in the p-block is not obvious in the later groups for two reasons. Firstly, for the elements in the top right of the p-block, electrons are too strongly held for high oxidation states to be readily formed, and secondly, the paucity of available chemical data for the radioactive 6p elements means that any tendency to form lower oxidation states is not apparent.

4.2 Element size and coordination numbers

The effect of atomic size is often underestimated or understated when considering the structures of inorganic molecules, but we should bear in mind that larger atoms can and do support larger coordination numbers. We shall see many examples of this as we look at the various classes of compound, but in addition to any structural effects, we should remember that, for a given coordination number, larger central atoms are also likely to be more reactive since there is easier access to the element centre.

4.3 Bond energies

4.3.1 Homonuclear bonds

Approximate bond energies, which are a measure of bond strength, for homonuclear single bonds for some of the s- and p-block elements are given in Table 4.4. We must exercise some care in using bond energy data since values can vary considerably depending on the compound involved (for example, the bond dissociation energy measured for the N–N bond in N_2H_4, N_2F_4 and N_2O_4 is 167, 88 and 57 kJ mol^{-1} respectively). In making detailed comparisons, therefore, it is necessary to compare similar compounds, but we are not concerned here so much with precise energies, but rather with reasonable values from which we can discern any trends which are apparent.

Some definition of bond energy is appropriate here. In H_2 for example, the bond dissociation energy ΔU_{el} is greater than the true bond dissociation energy ΔU_0 by an amount corresponding to the zero point energy even at 0 K (26 kJ mol^{-1} for H_2). The value at room temperature, ΔU_{298}, is slightly different again since translational, rotational and vibrational energy must taken into account and the dissociation enthalpy, ΔH_{298} includes a term relating to the work done to maintain constant pressure. ΔU_0, ΔU_{298} and ΔH_{298} are usually quite similar in magnitude and the latter is available to direct experimental measurement. We will mostly use ΔU_0 values.

Firstly, it is clear that, in general, bonds become weaker as the groups are descended. This is simply a result of the fact that the atoms are becoming larger and therefore the orbitals are also becoming larger and more diffuse to the point where overlap is poor and bonds are correspondingly weak.

An obvious exception to this generalization occurs for the first row elements of groups 15, 16 and 17. Note that whilst B–B and C–C bonds are the strongest of the group, the energies of the N–N, O–O and F–F single

Note the strength of the H–H bond.

bonds are unusually weak. The simplest explanation for this 'anomaly', as it is often called, has to do with the presence of lone pairs and the small size of the atoms concerned. Thus, each nitrogen, oxygen and fluorine atom has one, two and three non-bonded or lone pairs of electrons respectively which are held relatively close to the nucleus (bond pairs are shared between two atoms) and this factor, coupled with the particularly small size of these atoms, leads to appreciable inter-electron repulsions between neighbouring atoms thereby destabilizing the bond. In other words, for sufficient orbital overlap to occur to form a strong element–element bond, the atoms would have to get so close together that substantial repulsions between their non-bonded electron pairs would result. In the cases of boron and carbon, all the electron pairs are bond pairs, so that inter-electron repulsions around the central atom are much less and the atoms are also larger, both factors accounting for the presence of much stronger bonds. For the second and subsequent row elements, the larger size of the atoms means that inter-lone pair repulsions are less important (although see later) with the result that second row element-element bonds are the strongest in the group for groups 15 - 17 with subsequent rows becoming weaker as expected.

Table 4.4 Homonuclear single bond energies for the s- and p-block elements (kJ mol^{-1})

H–H 432						
Li–Li 105	Be–Be (208)	B–B 293	C–C 346	N–N 167	O–O 142	F–F 155
Na–Na 72	Mg–Mg (129)	Al–Al —	Si–Si 222	P–P 201	S–S 226	Cl–Cl 240
K–K 49	Ca–Ca (105)	Ga–Ga 115	Ge–Ge 188	As–As 146	Se–Se 172	Br–Br 190
Rb–Rb 45	Sr–Sr (84)	In–In 100	Sn–Sn 146	Sb–Sb 121	Te–Te (126)	I–I 149

Values in Table 4.4, and in subsequent Tables containing bond energy data, are taken from Huheey, Keiter and Keiter (1993). There is little data available for the 6s and 6p elements and aluminium. Values for the N–N and O–O bonds are estimated values derived from thermochemical cycles and values in parentheses are calculated values. No indication of errors is given which are substantial in some cases.

An additional part of the explanation of, for example, the weakness of the F–F bond in F_2 derives from the presence of filled antibonding F–F π^* orbitals as shown in Fig. 4.1 which is a simplified molecular orbital diagram for F_2. Remember that a 2-orbital-4-electron interaction is destabilizing so that the filled F_2 π^* orbitals do more than just 'cancel out' the filled π orbitals. For a full discussion of molecular orbitals, see Winter (1993).

Trends in bond energy across a period are somewhat less marked but it is clear that, in general, there is a significant increase in bond strength as we move from group 1 to group 14 with a slight decrease at group 15 followed by a roughly constant or modest increase in bond energy as we go from group 15 to group 17; for the 2p elements, the drop at group 15 (i.e. nitrogen) is considerable and the O–O and F–F bond strengths quoted are slightly lower than that for the N–N bond. The trend from group 1 to group 14 results from the increasing electronegativity of the atoms since the more electropositive atoms to the left of the periodic table, when mutually bonded, will tend to have destabilising, adjacent δ^+ charges. On moving from group

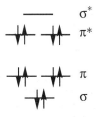

Fig. 4.1 A simplified molecular orbital energy level diagram derived from overlap of the 2p orbitals showing the σ and π molecular orbitals for F_2.

For example, in a compound $R_2Al–AlR_2$, the polarization will be $R_2Al(\delta^+)–(\delta^+)AlR_2$ since R is likely to be more electronegative than Al.

14 to group 15, the elements now have lone pairs (at least for the trivalent state), and the mutually repulsive interaction between these electron pairs across a bond, as discussed above for N–N, O–O and F–F bonds where it is most pronounced, is the main reason for the observed drop in bond energy.

4.3.2 Heteronuclear bonds

If we now look at heteronuclear bonds, it will become apparent that these are generally much stronger than the homonuclear bonds we have just considered, as is evident from the bond energies for the s- and p-block element fluorides displayed in Table 4.5.

Table 4.5 Bond energies for the s- and p-block elements fluorides (kJ mol^{-1})

H–F						
565						

Li–F	Be–F	B–F	C–F	N–F	O–F	F–F
573	632	613	485	283*	190*	155
Na–F	Mg–F	Al–F	Si–F	P–F	S–F	Cl–F
477	513	583	565	490*	284	142*
K–F	Ca–F	Ga–F	Ge–F	As–F	Se–F	Br–F
490	550	469	452	406	285	187*
Rb–F	Sr–F	In–F	Sn–F	Sb–F	Te–F	I–F
490	553	444	414	402	329	231
Cs–F	Ba–F	Tl–F	Pb–F	Bi–F	Po–F	At–F
502	578	439*	331	297	—	—

The bond energies quoted in Table 4.5 are for the elements in their group oxidation states except for those indicated with an asterisk. As with Table 4.4, errors are not given but may be considerable (up to ± 50 kJ mol^{-1}.

A general explanation as to the disparity between heteronuclear and homonuclear bond energies was offered by Pauling who ascribed it to the difference in electronegativity of the elements involved; indeed the concept of electronegativity largely evolved from a consideration of the differences between the strengths of homonuclear vs. heteronuclear bonds. Thus, any difference in electronegativity of the elements forming a bond will give rise to a certain degree of ionicity or charge separation (δ^+–δ^-) which will lead to an additional electrostatic attraction over and above that resulting from the covalent bond itself. Pauling proposed an equation to calculate the bond dissociation energy (D, in kJ mol^{-1}) of a heteronuclear bond based on the values of the respective homonuclear bonds and the square of the electronegativity difference between the atoms (Eqn 4.2).

In Eqn. 4.2, Pauling originally gave a proportionality constant of 1 where D was quoted in eV. To convert values of $\chi_A - \chi_B$ in eV to kcal mol^{-1}, a proportionality constant of 23 was later used but to convert to modern units of kJ mol^{-1}, the constant should be 100.

$$D_{(A-B)} = \frac{1}{2}[D_{(A-A)} + D_{(B-B)}] + 100(\chi_A - \chi_B)^2 \qquad (4.2)$$

Whilst Eqn 4.2 is only approximate, it is a useful guide to the strength of heteronuclear bonds and provides a simple rationale for the high bond energies to fluorine in particular, although other factors such as π-bonding effects are also important as we shall see.

Although a lattice energy and a covalent bond energy seem rather different, they are both a measure of the binding energy between atoms so there is no problem in discussing them together.

If we look first at the s-block elements, it is clear that the bond energies, or, to be more precise, lattice energies since we are dealing here with ionic compounds (see later), are all round about 500 - 600 kJ mol^{-1} which are very considerable values. As we descend groups 1 and 2 no particular trends are apparent, and the precise values will in any case depend on the details of the ionic crystal structure and are themselves subject to error, but it is clear that the bond energies for the group 2 fluorides are higher than those for the corresponding group 1 fluorides which results from the presence of 2+ cations in the case of the former; lattice energies are proportional to the charge of the ions.

In group 13, the values quoted in Table 4.5 are for compounds in the trivalent state, with the exception of Tl–F the value for which is derived from thallium monofluoride (bond energies vary with oxidation state, sometimes by a considerable amount, as we shall see), and show a general decrease as we descend the group. This does not mirror the trends in electronegativity whereby we might expect, for example, that the Al–F bond energy would be greater than that for a B–F bond, but we must recognise that size will also be important here and since aluminium is bigger than boron, this will work to weaken the Al–F bond relative to the B–F bond. Furthermore, since trivalent boron and aluminium compounds often have a vacant orbital, the possibility of B–F and Al–F π-bonding exists with the former expected to be considerably stronger for reasons discussed in the following section. Remember that Eqn. 4.2 tells us how much stronger an A–B bond will be in relation to the mean of A–A and B–B bond energies; it does not imply that bond energies are determined simply by element electronegativity differences.

In group 14, the general trend is the same as for group 13 for much the same reasons and for the heavier elements, it would appear that size considerations are probably more important than electronegativity differences such that bond strengths decrease down a group. However, the important difference in this group compared to group 13 is that the C–F bond is considerably weaker than the Si–F bond (c.f. B–F > Al–F). This is clearly not to be expected on the basis of element size and undoubtedly reflects the greater electronegativity difference between Si and F vs. C and F. Moreover, with silicon there is also the possibility of F→Si π-bonding due to the availability of low energy vacant orbitals on silicon, analogues of which are not present for carbon; the nature of these orbitals will be addressed in Chapter 6.

A situation where size and electronegativity difference work together can be seen by considering the carbon to halide vs. the silicon to halide bond energies shown in Table 4.6. Thus, there is a progression to weaker bonds as the halogen group is descended since the size of the halogen is increasing (resulting in poorer overlap) and the electronegativity difference is decreasing. Note, however, that in all cases, the Si–X (X = halogen) bonds are stronger than the C–X bonds for a given X as discussed above for the specific case of the fluorides.

These large bond energies for the ionic s-block fluorides coupled with the three-dimensional bonding present in ionic solids accounts in large part for the characteristic features of ionic materials such as hardness and high melting points. Note that the lattice energy ΔH_{latt} is proportional to z^+z^-/r^+r^- where z is the ionic charge and r is the ionic radius.

The comments here on π-bonding, i.e. B–F > Al–F but Si–F > C–F may appear confusing at this stage. Where a primary valence orbital is available (as with B and Al), π-bonding is expected to be strongest for the 2p element but where orbitals of a different type (generally vacant d orbitals or σ^*-orbitals; see Chapter 6) are present, π-bonding is generally greater for the 3p element.

Table 4.6 Heteronuclear carbon and silicon to halogen bond energies (kJ mol^{-1})

C–F	485	C–Cl	327	C–Br	285	C–I	213
Si–F	565	Si–Cl	381	Si–Br	310	Si–I	234

A similar situation is found for the oxides where Si–O (452) and P–O (335) single bonds are found to be much stronger than C–O (358) and N–O (201) single bonds respectively.

In group 15, the situation with the element fluorides is the same as that for group 14 but we should note that the N–F bond is much weaker than the P–F bond. This is a consequence of both N and F being small elements and having lone pairs, the effect of which was discussed in the case of homonuclear bonds, but we should note that the N–F bond is a lot stronger than either the N–N bond (167) or the F–F bond (155) which clearly demonstrates the effect of electronegativity difference. This feature is reflected in the chemistry of the nitrogen halides wherein NF_3 is a thermodynamically stable and unreactive molecule in complete contrast to the heavier congener, NCl_3, which is extremely unstable and reactive due largely to the small N–Cl bond energy (190 kJ mol^{-1}) which stems from the similar electronegativities of nitrogen and chlorine; P–H, C–I and S–I bonds tend to be rather weak for similar reasons. Bromides and iodides of nitrogen are even more unstable which also reflects the general trend to weaker bonds as a group is descended.

In groups 16 and 17, the O–F and F–F bonds are very weak (small atoms, lone pairs and small or zero electronegativity difference) but there is now a tendency for the element to fluorine bond energies to increase as we descend the group. This observation would indicate that it is now probably electronegativity differences which are dominant rather than size differences which is perhaps not unexpected since the elements of these groups are themselves generally smaller than those of the preceding groups such that the size mismatch between them and fluorine is less marked. Alternatively, it may be that with sulfur and chlorine, these atoms are sufficiently small for lone pair-lone pair repulsions across the S–F and Cl–F bonds to be important whereas with tellurium and iodine, for example, such a factor is much less dominant.

As a final point regarding Table 4.5, we should note that, in considering trends across a period, there is, in all cases, a marked decrease in the E–F bond energy as we go from left to right as would be expected from electronegativity considerations.

Before we leave a discussion of the strengths of heteronuclear bonds, we should consider the case of the lead(IV) chlorides and alkyls, described in section 4.1, since there are a number of important general points which emerge.

Recall that for lead chlorides, $PbCl_2$ is more stable than $PbCl_4$ whereas for the alkyls, $PbEt_4$ is more stable than $PbEt_2$. If we consider first the former observation, part of the reason is due to the relative weakness of the Pb(IV)–Cl bond and this highlights a general point concerning the bond strengths of lead(II) vs. lead(IV) halides, since it is found that Pb–X bonds in lead(II) halides are generally about 50–60 kJ mol^{-1} stronger than Pb–X bonds in lead(IV) halides. Moreover, the same is true for the respective +2 and +4 states of tin and germanium and also for indium where the In–X bond strengths for indium(I) halides may be as much as 100 kJ mol^{-1} greater than

The value quoted in Table 4.5 for the Cl–F bond (142) is actually lower than that for F_2 (155) but the value for chlorine in this case is derived from the molecule ClF_5. A better comparison with F_2 is the Cl–F molecule for which the Cl–F bond energy is 249 kJ mol^{-1}, i.e. considerably greater.

Rather surprisingly, perhaps, Pb(IV)–X bonds, although weaker than Pb(II)–X bonds, are significantly shorter which is due to the smaller size of the Pb(IV) centre and reflects the large size difference between the 6s and 6p orbitals.

those for indium(III) halides (the situation is similar for gallium and thallium).

Clearly, the involvement of the s^2 electron pair in the higher valent state leads to a considerable weakening of the bonds and this can be traced to the difference in the radial extension of the s and p orbitals of the heavier elements mentioned briefly in Chapter 1. The result of the markedly different sizes of the s and p orbitals for the heavier elements is that sp hybridization is less efficient and covalent bonding involving an sp hybridized state is correspondingly weaker; in the +2 state, bonding can occur using only p-orbitals. This is, of course, the same argument as was presented above in our discussion of the inert pair effect.

With the above points in mind, we should anticipate that electronegative substituents will result in particularly weak covalent bonding in situations where sp hybridization is important, since the high partial positive charge induced at the central element centre will tend to increase the difference in radial extension between the s and p orbitals making sp hybridization even less efficient. This is undoubtedly an important part of the reason for the instability of the halides of lead(IV) described above and should be most marked in the case of the fluoride, but we should expect that any weakening of the covalent part is likely to be offset by a strong ionic contribution to the bond energy which would account for the stabilty of PbF_4 vs. the heavier halides. This is, in fact, a general observation for the heavier p-block elements; high oxidation states involving removal of the s^2 pair are observed for fluorides but are much less common for the heavier halides.

In considering the case of the lead alkyls, the much lower electronegativity of carbon will not result in differentially much weaker bonds for the +4 oxidation state and there is likely to be much less of a difference in the Pb–C bond energies for the lead(II) and lead(IV) alkyls. As a consequence, lead will tend to form four bonds rather than two which is evident from the calculated exothermicity of the disproportionation reaction shown in Eqn 4.3.

$$2 \ Pb(II)Et_2 \quad \rightarrow \quad Pb(0) \quad + \quad Pb(IV)Et_4 \qquad (4.3)$$

The increasingly disparate size of the valence s and p orbitals as the group is descended is illustrated with the following values for the group 14 elements.

	$r_{max} \ ns$ (pm)	$r_{max} \ np$ (pm)
C	65	64
Si	95	115
Ge	95	119
Sn	110	137
Pb	107	140

There are other reasons for the instability of the heavier halides as we have previously mentioned, notably the increasing size of the halide and the progressive ease of oxidation by a high oxidation state element centre.

4.3.3 Multiple bonds

Finally in this section we should consider the trends in the strengths of multiple bonds. Multiple bonding is common amongst the first row (2p) elements and is a very important aspect of their chemistry. It is very much less common amongst the second and subsequent row elements, however, and we can begin to appreciate the reason for this from a consideration of the various bond energies associated with N_2 and P_2 fragments shown in Table 4.7.

Table 4.7 Single and multiple bond energies for dinitrogen and diphosphorus units (kJ mol^{-1})

N–N	167	P–P	201
N=N	418	P=P	310
N≡N	942	P≡P	481

Several points are important. Firstly, note that the N≡N triple bond is extremely strong (only the C≡O bond in carbon monoxide is stronger) and almost twice as strong as the P≡P triple bond. Secondly, although the N=N double bond is stronger than the P=P double bond, the more important point is that the former is considerably more than twice the strength of the N–N single bond whereas the reverse is true for the latter vs. the P–P single bond. This second observation can be seen as a result of the particular weakness of the N–N single bond, for reasons already discussed, with the important consequence that in the chemistry of these elements, phosphorus prefers to form two single bonds rather than one double bond, whilst the opposite is true for nitrogen.

In seeking an explanation for the varying stability of $p\pi$–$p\pi$ multiple bonds between the p-block elements, we must look further at the strengths of the various bonds involved and an illuminating way in which this can be done (specifically for double bonds in this case) is to examine the σ and π bond increments for particular pairs of atoms. Table 4.8 lists such increments for several pairs of elements (nitrogen and phosphorus included) and the values are defined such that the first value for each particular bond is that for a typical element–element single or σ bond; for example the value of 335 kJ mol^{-1} for carbon is that of the C–C bond in ethane. The second value is the difference in energy between a double, or $\sigma + \pi$ bond and that of the single bond given by the first value and this can be taken as a measure of the strength of the π bond. For carbon, the difference is that between the bond energy of the C=C double bond in ethene (630 kJ mol^{-1}) and the single C–C bond in ethane.

Clearly the σ bond in ethane is not the same as the σ bond in ethene but the differences are likely to be small; moreover, it is broad comparisons between different types of bonds that are important rather than a means of quantitatively predicting particular bond energies.

Values in Table 4.8 are taken from Kutzelnigg (*Angew.Chem.*, (1984), **23**, 272). Note that there are slight differences between some of the values in Table 4.8 and those in Table 4.4. This reflects the use of different sources but, as was mentioned earlier, there are no unique values for bond energies; they depend to some extent on the molecule. However, these differences are not important; it is the trends which are important rather than absolute values.

Table 4.8. Approximate bond increments (kJ mol^{-1}) for σ/π bonds for various element–element pairs

C–C	N–N	O–O	C–O	N–O
335/295	160/395	145/350	335/380	190/370
Si–Si	P–P	S–S	Si–O	P–O
195/120	200/145	270/155	420/170	335/150
Ge–Ge	As–As	Se–Se	C–S	S–O
165/110	175/120	210/125	280/265	275/250

If we first consider homonuclear bonds, an important point to emerge from Table 4.8 is that the π bond increment for the first row or 2p elements is considerably greater than that for the elements of the second and subsequent rows. In other words, π bond strengths are much larger for the first row elements and by a factor of between two and three. The standard explanation for this feature is that the second and subsequent row elements, being larger, have correspondingly larger and therefore more diffuse orbitals. Overlap between these bigger orbitals is poorer which results in weaker bonds, particularly in the case of π bonds in which the atomic orbital overlap is side-on rather than the head-on overlap of σ orbitals. When considering the disparity in the prevalence of π bonding between the 2p elements and those of subsequent rows however, another aspect which we must consider is the weakness of the N–N and O–O single bonds (like the F–F bond). As a result,

P–P and S–S single bonds are considerably stronger than N–N and O–O bonds respectively. In considering both of these factors, it is therefore apparent that for groups 15 and 16, as the values in Table 4.8 clearly indicate, the combination of a $\sigma + \pi$ bond is more stable than two σ bonds for the first row elements whereas the opposite is true for the elements of the second and subsequent rows.

In group 14 the comparison is a little different in that the strength of the C–C single bond is greater than that of the Si–Si bond since there are no non-bonded electron pairs associated with the carbon atoms. The absence of an extensive chemistry of multiply bonded silicon compounds is due, as before, to the weakness of π-bonding for this element (two σ bonds are stronger than one σ and one π bond) and it is noteworthy that Si–Si π-bonding is the weakest of the 3p elements consistent with the larger size of the silicon atom (vs. P or S). The occurance of π-bonding in carbon chemistry, however, is not associated with any weakness of the C–C σ bond but we may note that much of the extensive and rather special organic chemistry of carbon is the result of the similarity in the strengths of the C–C σ and π bonds. With this final point in mind, it is also clear from the values for group 14 in Table 4.8 that, in absence of factors which weaken the first row σ bonds, the ratio of the σ/π bond strengths drops markedly between the first and second row (0.88 vs. 0.61) but then remains nearly constant for the second and third row (0.61 vs. 0.66).

An examination of σ and π increments for heteronuclear bonds is also instructive. For example, Table 4.8 reveals that the π bond increment is large where both elements are from the first row although the σ bond strengths are also large where one of the elements is carbon. Heteronuclear bonds where a second row element is involved, for example Si–O and P–O, are similar to the homonuclear second row element cases in having a small π increment but we should note the rather special case of sulfur wherein the σ and π increments for both C–S and S–O bonds are about equal. This is mainly the result of a large π increment (as opposed to weak σ-bonding, c.f. C–C) and in this sense, sulfur is acting (as far as π-bonding is concerned) more like a first row element. This is undoubtedly due to the relatively small size of the sulfur atom and the consequent better π overlap, and it is therefore not surprising that pπ–pπ multiple bonding is important in the chemistry of this element. We might therefore imagine that the diatomic molecule S_2 would be a stable form of sulfur and indeed, the strength of the S=S double bond in S_2 (425 kJ mol^{-1}) is only slightly less than that of the O=O double bond in O_2 (495 kJ mol^{-1}). Note, however, that the large π increment for oxygen favours a doubly bonded structure over one with single bonds, whereas the reverse in true in the case of sulfur. This latter point is reflected in the exothermicity of the reaction shown in Eqn 4.4.

$$4\,S_2 \rightarrow S_8$$
$$\Delta H = -520 \text{ kJ mol}^{-1} \qquad (4.4)$$

4.3.4 Bonds to p-block elements vs. d- and f-block elements

Before we leave this section, it is worth noting a further difference between p-block chemistry and that of the d-block elements. As we have seen, bond energies generally decrease as groups in the p-block are descended, but the opposite is often true for the d-block elements with bonds to third row elements tending to be the strongest. An explanation for this derives from that given in Section 4.1 concerning trends in oxidation number. Thus the relativistic stabilization and contraction of the s and p orbitals for the third row or 5d elements leaves the d orbitals more exposed and so overlap is better and bonds are stronger. A similar situation is found for the f-block elements where covalent bonding is much more prevalent in the chemistry of the actinides.

This is well illustrated in the case of the tetrafluorides of titanium, zirconium and hafnium. Recall that in group 14, the promotion energies are increasing down the group and the bond strengths are decreasing which leads to the appearance of an inert pair effect. For the group 4 elements, the promotion energies (s^2d^2 to s^1d^3 in this case) are also largest for the heaviest element (Ti, 79; Zr, 58; Hf, 169 kJ mol^{-1}) but the M–F bond energies increase down the group (Ti–F, 588; Zr–F, 647; Hf–F, 676 kJ mol^{-1}).

4.4 The van Arkel-Ketelaar triangle

Before looking at the compounds of the s- and p-block elements in detail, it will be useful to consider the general types of compound we will expect and this, as we shall see, can often be rationalized on the basis of electronegativities, χ, and electronegativity differences, $\Delta\chi$. Thus compounds with a large $\Delta\chi$ will tend to be ionic and those with medium $\Delta\chi$ are likely to form polar covalent bonds and therefore have polymeric or macromolecular structures. For compounds with a small $\Delta\chi$, the type of compound will depend on whether the elements have a low or high electronegativity. In the former case, the compounds are likely to be metallic for much the same reason that elements with low electronegativity are metals. In the latter case, we shall expect relatively non-polar molecular species.

A useful way to visualize these ideas is with the so-called van Arkel-Ketelaar or element triangle, shown in Fig. 4.2, in which the vertices are labelled metallic, covalent and ionic in such a way that the vertical and horizontal axes represent electronegativity difference, $\Delta\chi$, and absolute electronegativity, $\Sigma\chi$, respectively and are defined such that, for a molecule AB, $\Delta\chi = \chi_B - \chi_A$ and $\Sigma\chi = (\chi_A + \chi_B)/2$.

The metallic–covalent edge of the van Arkel-Ketelaar triangle comprises the elements themselves, since $\Delta\chi$ is zero, and ranges from metals on the left to covalent species such as F_2 on the right for reasons addressed in Section 2.5 in Chapter 2. An important sub-category in this section is the metalloids such as boron and graphitic carbon, but also including silicon, germanium, arsenic and antimony. In all cases, covalent bonding resulting from overlap of orbitals is a characteristic feature and it has been suggested by Allen and Burdett that the specific term 'metallic bond' actually be dropped from inorganic texts since whilst there is certainly a metallic state with defining properties such as electrical conductivity, the bonding in metals is covalent in nature and does not differ in any fundamental way from other types of covalent bonding.

The ionic–covalent edge is characterized by decreasing $\Delta\chi$ which is well illustrated with the fluorides of the elements. Thus, CsF and LiF are ionic with large $\Delta\chi$ whereas NF_3 and CF_4 both have small $\Delta\chi$ and are covalent; compounds with intermediate $\Delta\chi$ such as BeF_2 are polymeric or macromolecular solids. A similar trend is seen with the element oxides and both classes are described in detail in the next chapter.

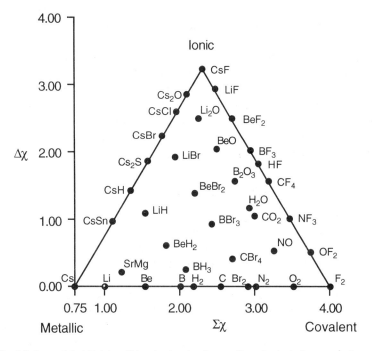

Fig. 4.2 is derived from semi-quantitative or quantitative treatments given by Allen (*J.Am.Chem.Soc.*, (1992), **114**, 1510) and Jensen (*J.Chem.Educ.*, (1995), **72**, 395) building on the original work of van Arkel and Ketelaar. Allen electronegativities have been used for the elements and compounds displayed.

All discussion is for compounds at atmospheric pressure; metallicity increases with pressure as noted previously for diiodine.

Fig. 4.2 A van Arkel-Ketalaar Triangle showing the position of selected compounds and elements.

The metallic–ionic edge is also characterized by increasing $\Delta\chi$ and is illustrated in Fig. 4.2 with compounds of caesium ranging from caesium itself through the metallic CsSn to the ionic compounds Cs_2O and CsF.

The examples looked at so far were chosen to lie along the edge of the triangle in order to illustrate the extremes of metallic, covalent and ionic bonding, but many (in fact, most) compounds will have values of $\Delta\chi$ and $\Sigma\chi$ which place them within the area of the triangle and they will therefore have chatacteristics of all three types of bonding. This is illustrated for a selection of compounds in Fig. 4.2 where any compound can be plotted as a function of $\Delta\chi$ and $\Sigma\chi$. Notice, for example, that as for the fluorides which lie along the ionic to covalent edge, the oxides and bromides lie along straight lines which are parallel to this edge but within the triangle since $\Delta\chi$ values are less in the latter two cases. Similarly, just as caesium compounds lie along the ionic to metallic edge, compounds of less electropositive lithium, beryllium and boron lie within the triangle on lines parallel to this ionic-metallic edge. Note that because of the way the diagram has been constructed, compounds AB are written with B more electronegative than A so that for any given class of compound where B remains the same and A is varied, such compounds will lie along a line parallel to the ionic-covalent edge. Likewise,

A much fuller account of bonding in terms of the element triangle can be found in Alcock (1990).

for AB where we look at a range of B for a given A, the compounds will lie on a line parallel to the ionic-metallic edge.

As we saw in Chapter 2 (see Fig. 2.6), the metalloid region for the elements is defined by a quite narrow region of electronegativity. With regard to compounds, a particularly important class is the binaries of elements in or near the metalloid region such as GaAs, InSb and SiC together with the binaries for which the element electronegativities are themselves outside the metalloid range, but where the average value, $\Sigma\chi$, lies within this range and where $\Delta\chi$ is sufficiently small for covalent rather than ionic bonding to be present; examples are AlP, GaP and InP. We can therefore define a metalloid region in the element triangle which is shown in Fig. 4.3. Similarly, we could define a metallic region, the usefulness of these regions being that for any given compound, we can plot its position in the triangle and have an immediate feel for its likely properties.

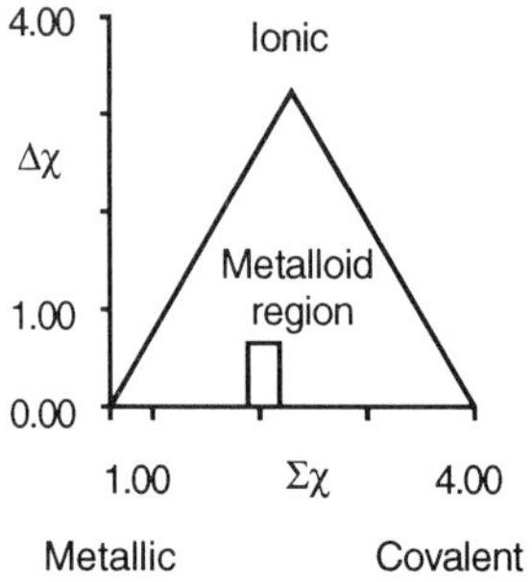

The treatment outlined above can be taken a step further if we recognise that in addition to metallic, covalent and ionic bonding, we should also consider van der Waals interactions between discrete covalent molecules in the solid state. Such consideration leads us to the so-called element tetrahedron shown in Fig. 4.4 in which each of the four vertices represents the four bonding types. We can take solid di-iodine as an example of a material close to the van der Waals vertex wherein the iodine atoms are linked by a covalent I–I bond whereas the I_2 molecules are held together by much weaker van der Waals interactions. We will not look at this element tetrahedron in any great detail (the ionic-covalent-metallic face is the same as the van Arkel-Ketelaar triangle, of course), but it is informative to comment on compounds which lie along the three edges which join at the van der Waals vertex.

A material mid-way along the van der Waals-covalent edge might be selenium wherein the intermolecular Se····Se interactions between the helical Se_n chains are sufficiently short to suggest the onset of some degree of covalency in addition to any van der Waals bonding (Se–Se = 2.37 Å, Se····Se = 3.44 Å). Gallium is a good example for a material part way along the van der Waals-metallic edge since gallium has definite metallic properties but the presence of discrete Ga_2 units (see Section 3.2) is suggestive of diatomic units linked by weaker bonding. Finally, as an example along the van der Waals-ionic edge, we might consider $AlBr_3$ which is a low-melting solid of close packed bromines with aluminium atoms in one sixth of the tetrahedral holes, and which readily sublimes to give Al_2Br_6 molecules.

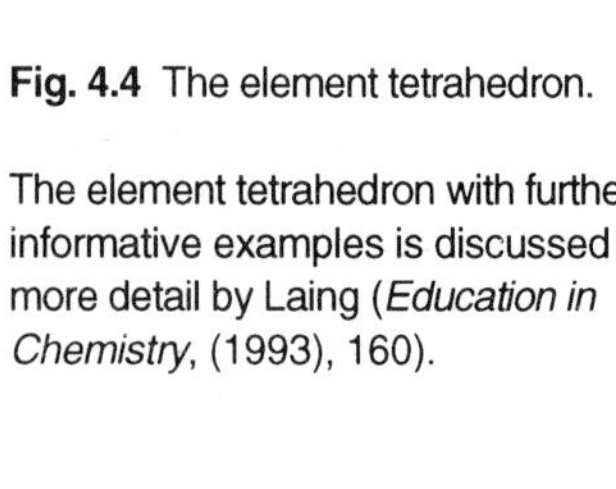

Having explored some general aspects, it is now time to look at s- and p-block element compounds and their properties in more detail.

Fig. 4.3 A simplified van Arkel-Ketelaar triangle showing the metalloid region defined by values of $\Delta\chi$ and $\Sigma\chi$.

Fig. 4.4 The element tetrahedron.

The element tetrahedron with further informative examples is discussed in more detail by Laing (*Education in Chemistry*, (1993), 160).

5 Compounds of the s- and p-block elements

5.1 Element halides

The halides are the first class of compound we shall look at in detail and it will soon be clear that it is a large and diverse group particularly in terms of the variety of structural types encountered. Many of the ideas which we develop here will also be of use when looking at other classes of compound and we shall start with a broad overview of the basic structural types.

We shall look first at the fluorides of the s- and p-block elements in their highest, or group, oxidation states; these are shown in Table 5.1 together with a classification based on the type of structure they adopt under ambient conditions.

Table 5.1 A structural classification of the s- and p-block element fluorides

1	2	13	14	15	16	17
LiF	BeF_2	BF_3	CF_4			
NaF	MgF_2	AlF_3	SiF_4	PF_5	SF_6	
KF	CaF_2	GaF_3	GeF_4	AsF_5	SeF_6	
RbF	SrF_2	InF_3	SnF_4	SbF_5	TeF_6	IF_7
CsF	BaF_2	TlF_3	PbF_4	BiF_5		

ionic   polymeric   molecular covalent

We note first that the formulae are of the form EX_n where n = 1–7 according to the group and that there is a definite progression from ionic materials in the bottom left of the table to molecular covalent species in the top right, with polymeric or macromolecular solids forming an approximately diagonal band in between. We can account for this general arrangement in a straightforward manner by considering the electronegativity differences, $\Delta\chi$, between the elements and we shall look at each in turn.

Fluorine is the most electronegative element in the periodic table (with the exception of some of the noble gases which are rather special cases). Thus, in forming complexes with the electropositive elements on the left of the periodic table, particularly those near the bottom left, there will be a considerable degree of charge transfer resulting in the formation of the negatively-charged fluoride anion, F^-, and a positively-charged element cation, E^+ (more generally E^{n+} where n is the number of valence electrons). The resulting materials formed are therefore ionic and adopt solid, three-

The fluorides of nitrogen, oxygen, krypton and xenon such as NF_3, OF_2, KrF_2 and XeF_6 are not shown in Table 5.1 since the elements in these compounds are not in their highest or group oxidation state.

A word of caution is necessary concerning electronegativities. In Chapter 2, we discussed electronegativity in some detail and quoted values derived from the work of Pauling and Allen. These values, especially from the latter, are neutral atom electronegativities and we can expect that the actual values will change according to the oxidation state of the element in a particular molecule. We will look at this matter again in more detail later, but it is reasonable to suppose that element electronegativity will increase with increasing oxidation state. Thus whilst tellurium and iodine have neutral atom electronegativities considerably less than fluorine, in the +6 and +7 oxidation states respectively, their electronegativities will be much more comparable.

dimensional structures, or lattices, in which cations are surrounded by anions and vice versa. The bonding is predominantly electrostatic in nature, the lattice being held together largely as a result of the opposite charges on the ions. In general, this type of bonding is fairly strong and ionic solids are characterized by having high melting and boiling points and by dissolving only in polar solvents such as water. Typical ionic structures are those of sodium fluoride and calcium difluoride shown in Fig. 5.1.

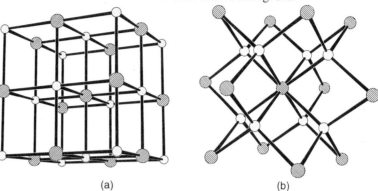

(a) (b)

Fig. 5.1 Diagramatic representations of the structures of (a) sodium fluoride, NaF, and (b) calcium difluoride, CaF_2.

If we look now at the molecular covalent species in the top right of Table 5.1, it is clear that all are compounds formed by elements with a similar electronegativity to fluorine.

The result is that there is only a small degree of charge transfer on bond formation and the element–fluorine bonds are relatively non-polar, covalent bonds. Thus, there is little or no aggregation between molecules and the compounds are typically gases, low boiling liquids or low melting solids which are soluble in non-polar solvents. Some examples are shown Fig. 5.2.

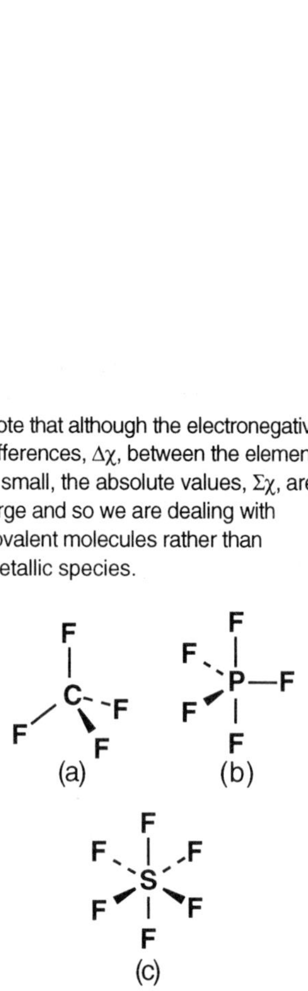

The intermediate class of compound, labelled polymeric in Table 5.1, consists of those in which the atoms have a medium difference in electronegativity. The difference is not sufficient for the formation of ions and ionic bonding, but the covalent bonds formed are quite polar with large partial positive and negative charges, i.e. δ^+ E–F δ^-. The effect of this in the solid state is that bridging interactions tend to occur between atoms of opposite charge leading to polymeric or macromolecular structures. Whilst aggregation or polymerization is a consequence of charge separation or polar bonds we should not assume that the resulting intermolecular interactions between the putative monomers are entirely electrostatic. All of the bonding is predominantly covalent and we can think of it as arising from the polymerization of monomers, a requirement for which is the presence of low-lying vacant orbitals. These may be vacant primary valence orbitals, as in group 13 chemistry, or other types of orbital the nature of which we will address in Chapter 7. An example is the structure of SbF_5 shown Fig. 5.3, which exists in a linear polymeric form in the liquid state and as tetramers in the solid.

Note that although the electronegativity differences, $\Delta\chi$, between the elements is small, the absolute values, $\Sigma\chi$, are large and so we are dealing with covalent molecules rather than metallic species.

(a) (b)

(c)

Fig. 5.2 Diagrams of the structures of (a) CF_4, (b) PF_5 and (c) SF_6.

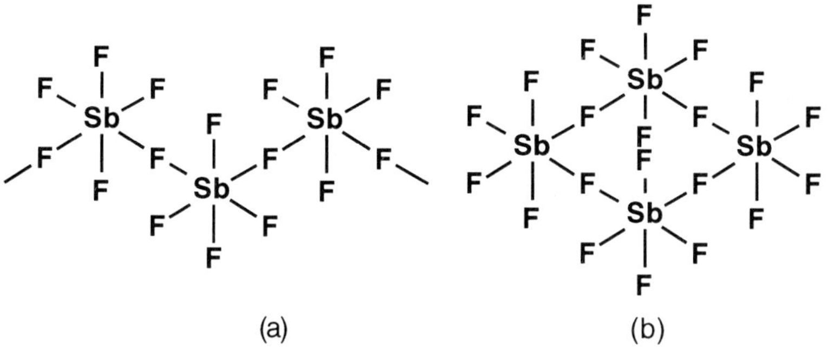

Fig. 5.3 Diagramatic representations of the structure of SbF$_5$ in (a) the liquid and (b) the solid state.

It should not be assumed that the above three classes of compound are completely distinct from each other since we cannot categorize bonds as simply ionic, polar covalent or covalent; all bonds lie on a continuum from highly ionic to purely covalent. There is certainly some overlap and occasional resulting ambiguity but the categories are useful nevertheless.

So far we have seen that the broad classification of the element fluorides, at least for elements in their highest oxidation states, can largely be rationalized on the basis of electronegativity differences, $\Delta\chi$. We shall now look at some of the lower oxidation state fluorides and see how these ideas can be developed to understand their structures as well.

There are no stable lower oxidation state fluorides in group 2, but in group 13 we encounter examples of boron(II) and thallium(I), B$_2$F$_4$ and TlF respectively. The former is molecular covalent and the latter ionic in accord with their electronegativity differences. Furthermore, it is interesting to compare the thallium halides TlF$_3$ and TlF. We should expect the former to be 'more covalent' than the latter, i.e. TlF should be the most ionic, since the Tl(III) cation will be much more polarizing than the Tl(I) cation and in this sense, the Tl(III) cation is more electronegative than the Tl(I) cation (see Section 2.5 in Chapter 2). Whilst this resulting difference in $\Delta\chi$ for Tl(III)F$_3$ and Tl(I)F (the latter being larger) is not obviously reflected in the structures and physical properties of the solids, it is noteworthy that TlF$_3$ hydrolyses readily in water to give Tl(OH)$_3$ and HF whereas TlF dissolves to give aquated ions, the latter being a feature typical of metals.

In group 14 the other common oxidation state is +2, at least for the heavier elements, and the element difluorides and tetrafluorides are interesting classes of compounds which offer a useful illustration of the concepts we have developed so far to rationalize structure. Recall (Table 5.1) that GeF$_4$ is molecular covalent whereas SnF$_4$ and PbF$_4$ are polymeric. We can partly rationalize this observation on the basis of electronegativity differences but we should also note that the larger tin and lead atoms are also better able to support larger coordination numbers (four for Ge in GeF$_4$ and six for Sn and Pb in SnF$_4$ and PbF$_4$).

GeF$_2$, in contrast to GeF$_4$, is polymeric and we can also use the idea of electronegativity differences to account for this observation (in part) if we consider that Ge(II) is likely to be less electronegative than Ge(IV) since, as

Electronegativity differences are not the only factor which determine the type of structure adopted. For example, aluminium trifluoride, AlF$_3$, is polymeric whereas silicon tetrafluoride, SiF$_4$, is a monomer, and yet aluminium and silicon have very similar electronegativities and therefore similar electronegativity differences with fluorine. The main reason why AlF$_3$ is polymeric is due to the presence of a vacant orbital on the aluminium centre (in the monomer) which results in the facile formation of extra bonds (in this case three extra bonds since in solid AlF$_3$ each aluminium is surrounded by an octahedron of fluorines; we may consider the bonding here to be multicentre covalent or approaching ionic). We shall encounter a rather similar situation with the boron hydrides.

A word should be said here about polarization and polarizability. Simple definitions will suffice in that a small highly charged cation will be strongly polarizing such that it will tend to strongly attract electrons towards itself, whereas a large, singly charged cation will be weakly polarizing. Conversely, a large anion will be easily polarized, i.e. its electron density will be easily distorted, whereas a small anion will be less easily polarized (these ideas will be relevant later when we look at acids and bases). In the case of ionic compounds, the combination of a strongly polarizing cation and an easily polarized anion leads to increasing covalency as illustrated by the examples below.

LiF > CsF

NaI > NaF

AlN > NaF

$\xrightarrow{\hspace{3cm}}$

More ionic

mentioned in general above, a Ge(IV) centre will be more polarizing, and hence more electronegative, than a Ge(II) centre. We commented on a similar situation for Tl(III) vs. Tl(I), but the consequences are more dramatic here since the polarity of the Ge–F bond in GeF_2 is now sufficiently large to result in a polymeric structure, whereas in GeF_4, the polarity is small enough for a covalent molecular structure to be favoured.

We should therefore expect that for the fluorides of a given element (and, as we shall see, element halides and oxides in general), there will be a trend from covalent to polymeric to ionic as the central element oxidation state, and hence electronegativity, decreases. This is illustrated not so much with SnF_2 which is also polymeric, but with PbF_2 which adopts typically ionic structures; the α form is analogous to $BaCl_2$ and the β form has the fluorite structure, the coordination numbers around the lead atom being nine and eight respectively.

Nevertheless, whilst electronegativity differences *are* a significant factor, we must not overstate their importance, or ignore other factors such as element size and the nature of any vacant orbitals which are present. Thus, GeF_2 is also more likely to be polymeric than GeF_4, since the former is only two-coordinate (as a monomer, and clearly coordinatively unsaturated) whereas the latter is four-coordinate. This factor coupled with the larger size of Ge(II) vs. Ge(IV) would enable Ge(II) to expand its coordination number, i.e. polymerize, more readily than Ge(IV). Furthermore, the presence of a vacant p orbital in monomeric GeF_2 would make this species particularly acidic and therefore prone to polymerisation; the analogy between AlF_3 and SiF_4 mentioned earlier is relevant here and we will return to this matter again in Chapter 7. Another factor to note is that whilst monomeric GeF_4 is tetrahedral and has no dipole moment, monomeric GeF_2, for which a bent structure is expected, would have a dipole moment which would facilitate aggregation.

In group 15 all elements have a stable trifluoride for which we see the expected progression in structural types. Thus NF_3, PF_3 and AsF_3 are all molecular whereas SbF_3 is polymeric. BiF_3 is best described as ionic and contrasts with the polymeric BiF_5, this observation again being consistent with the lower electronegativity and polarizing power of Bi(III) vs. Bi(V).

In group 16 there are three common oxidation states and in addition to SF_6, the fluorides SF_2, S_2F_2 (two isomers), S_2F_4, SF_4 and S_2F_{10} are also known, all of which are molecular. The chemistry of selenium is more limited and, besides SeF_6, only SeF_4 is stable. Selenium tetrafluoride is molecular and has the same structure as SF_4 but the analogous tellurium compound, TeF_4, is polymeric with a structure shown in Fig. 5.4.

In Table 5.1, the only fluoride of group 17 was IF_7 since this is the only known neutral interhalogen compound with a central element in the formal +7 oxidation state; there are no group 18 halides in the +8 oxidation state. We should note, however, that a number of neutral lower oxidation state group 17 and 18 fluorides are known. Examples are EF_3 and EF_5 (E = Cl, Br, I) and XeF_n (n = 2, 4, 6) all of which are molecular; ClF_3 is shown in Fig. 5.5.

Note that coordination numbers are generally lower for lower oxidation states. In polymeric GeF_2, the germanium atom is only three-coordinate whilst in monomeric GeF_4 it is four-coordinate. This is a result of the presence of a lone pair of electrons in the former which occupies one of the valence orbitals leaving only three for bonding (formally two normal Ge–F bonds and one dative F → Ge bond in this case).

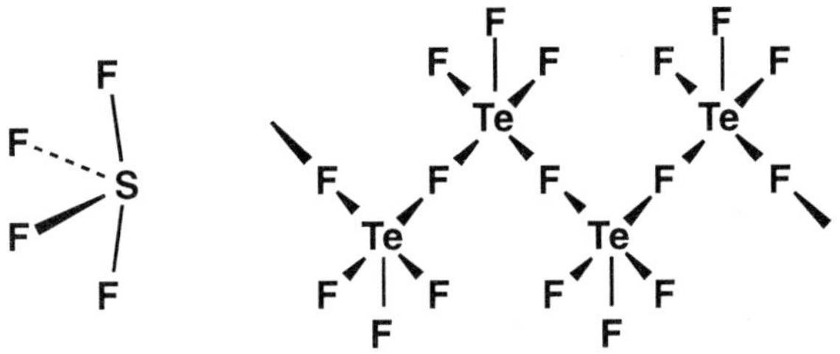

Fig. 5.4 The structure of (a) SF$_4$ and (b) part of the solid state structure of TeF$_4$.

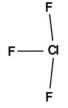

Fig. 5.5 The structure of ClF$_3$.

Before moving on to look at other halide compounds we should again stress the tendency of elements to form polymeric fluorides as groups are descended which results from increasing electronegativity differences and from the fact that element sizes are increasing.

Let us now consider compounds formed by the other halides. We should expect, since chlorine, bromine and iodine are less electronegative than fluorine, that there will be fewer ionic compounds formed and that there will also be a trend towards compounds with intermetallic or metallic alloy properties as we shall see with some of the iodides.

Table 5.2 A structural classification of the s- and p-block element chlorides

1	2	13	14	15	16	17
LiCl	BeCl$_2$	BCl$_3$	CCl$_4$	NCl$_3$		
NaCl	MgCl$_2$	AlCl$_3$	SiCl$_4$	PCl$_5$ PCl$_3$	SCl$_2$	
KCl	CaCl$_2$	GaCl$_3$	GeCl$_4$	AsCl$_5$ AsCl$_3$	SeCl$_4$	BrCl
RbCl	SrCl$_2$	InCl$_3$ InCl	SnCl$_4$ SnCl$_2$	SbCl$_5$ SbCl$_3$	TeCl$_4$	ICl$_3$
CsCl	BaCl$_2$	TlCl TlCl$_3$	PbCl$_2$	BiCl$_3$		

| Ionic | | | Polymeric | Molecular covalent | Polymeric | |

Only selected chlorides are shown in Table 5.2. Chlorides in the group oxidation state are indicated where these are known together with some common lower oxidation state species, but a number of other low oxidation state or sub-chlorides such as B$_8$Cl$_8$ and related compounds are not shown.

Table 5.2 lists the common chlorides of the s- and p-block elements although in contrast to Table 5.1, the lower oxidation state complexes are included as well, since the number of known chlorides of the elements in their group oxidation state is not large particularly for the right-hand side elements. This trend is even more marked for the bromides and iodides and is a result of the fact that the heavier halides become progressively easier to oxidize and are therefore correspondingly less able to stabilize high formal oxidation states.

The chlorides of groups 1 and 2 are ionic as expected with the exception of $BeCl_2$ which has a chloride-bridged, linear polymeric structure as shown in Fig. 5.6.

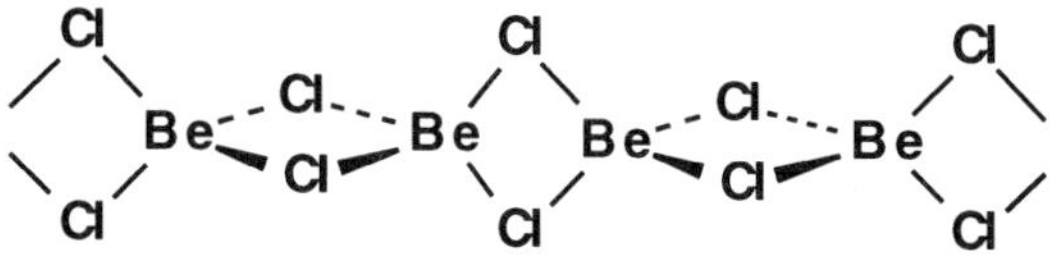

Fig. 5.6 Part of the structure of $BeCl_2$.

In group 13, $TlCl_3$ can be considered as ionic, but we should also include in this classification the monochlorides of thallium and indium. Both have typically ionic-type structures; TlCl has the CsCl structure and InCl has a distorted NaCl structure. Note that in the former structure the Tl(I) cation is eight-coordinate whereas in InCl the indium is six-coordinate. These differences undoubtedly reflect the larger size of Tl(I) vs. In(I).

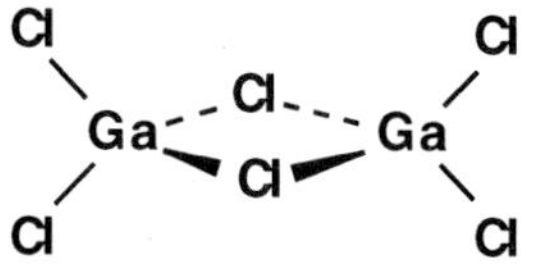

The solid-state structures of $AlCl_3$, $GaCl_3$ and $InCl_3$ are probably best described as polymeric. This is certainly the case for $AlCl_3$ which adopts a layer structure with six-coordinate aluminium centres, but for the trichlorides of gallium and indium the description is a little strained in that these compounds exist as dimers in the solid state (Fig. 5.7). In contrast to the heavier congeners, BCl_3 is monomeric probably as a result of appreciable B–Cl π-bonding. We noted in the previous chapter that π bond energies are often substantial where first row elements are involved and this is the likely explanation for the monomeric character of all the boron trihalides. The polymeric nature of the halides of the heavier congeners reflects a tendency for single bond formation, achieved in this instance by halide bridge formation, and the larger size of the elements means they are more able to support a larger coordination number.

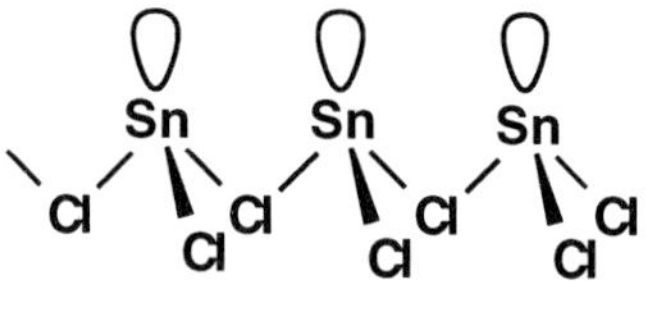

In group 14, the tetrachlorides of carbon, silicon, germanium and tin are all molecular whereas the dichlorides of tin and lead ($PbCl_4$ is not stable) are polymeric. A view of the polymeric structure of $SnCl_2$, illustrating the three-coordination at tin and the presence of the lone pairs, is shown in Fig. 5.8 and we note again the change from molecular to polymeric resulting from the change in oxidation state of the central element. Recall that a similar transition from one structural type to another was observed for GeF_4 vs. GeF_2.

All of the chlorides of group 15 can be described as molecular but there is an interesting point to make regarding PCl_5 which is ionic in the solid state, not as a result of the formation of naked phosphorus cations and chloride anions, but due to chloride exchange which affords the salt $[PCl_4]^+ [PCl_6]^-$. A similar situation is found for the phosphorus bromide PBr_5 which exists as $[PBr_4]^+ Br^-$. This ionization is not so much a result of particularly polar P–Cl or P–Br bonds but rather, is probably due to the significant lattice energy which results from the formation of an ionic solid, this being

Fig. 5.7 The structure of Ga_2Cl_6.

Fig. 5.8 Part of the solid state structure of $SnCl_2$.

considerably larger than any van der Waals interaction between neutral molecules.

In groups 16 and 17 the chlorides SCl_2 and BrCl are molecular whilst $SeCl_4$, $TeCl_4$ and ICl_3 are polymeric. For the latter three compounds, oligomeric might be a better description since the former two are tetrameric and ICl_3 exists as the dimer I_2Cl_6 the structures of which are shown in Fig. 5.9.

We will not dwell on the structures of the element bromides and iodides since the same general rules which we have encountered for fluorides and chlorides apply equally well for the heavier congeners. There is a general trend towards less ionic structures, as we might expect, and the number of stable compounds in the group oxidation state becomes progressively less, due partly to the fact that bromide and iodide are more readily oxidized and for other reasons discussed in Chapter 4.

5.2 Element oxides

A classification of the s- and p-block element oxides according to whether they are ionic, polymeric or molecular covalent is presented in Table 5.3 and the similarity to Tables 5.1 and 5.2, is obvious.

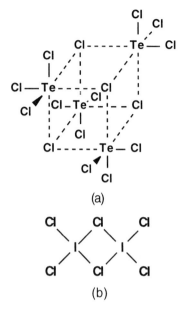

(a)

(b)

Fig. 5.9 The solid-state structures of (a) $TeCl_4$ and (b) ICl_3 (I_2Cl_6).

Only a selection of element oxides are shown in this Table (as was also done for the chlorides in Table 5.2). Examples of oxides not shown are the superoxides and peroxides of the group 1 elements, EO_2 and E_2O_2.

Table 5.3 A structural classification of the s- and p-block element oxides

1	2	13	14	15	16	17	18
Li_2O	BeO	B_2O_3	CO CO_2	NO N_2O N_2O_3 N_2O_4 N_2O_5	O_2 O_3	F_2O F_2O_2	
Na_2O	MgO	Al_2O_3	SiO_2	P_4O_6 P_4O_{10}	SO_2 SO_3	Cl_2O ClO_2 Cl_2O_7	
K_2O	CaO	Ga_2O_3	GeO_2	As_4O_6 As_2O_5	SeO_2 SeO_3	Br_2O BrO_2	
Rb_2O	SrO	In_2O_3	SnO_2 SnO	Sb_4O_6 Sb_2O_5	TeO_2 TeO_3	I_2O_4 I_2O_5 I_2O_9	XeO_3 XeO_4
	BaO	Tl_2O_3 Tl_2O	PbO_2 PbO	Bi_2O_3	$\boxed{PoO_2}$		
Ionic				Polymeric			Molecular covalent

Polonium dioxide is probably best described as ionic and so is drawn in a box in the section for polymeric compounds.

As we would expect, the left-hand element oxides are predominantly ionic due to the large electronegativity difference with oxygen, but we will not comment on these compounds or review the structures in any detail except to point out that the structures are typical of ionic materials. We should be aware, however, that there are occasions when the classification, particularly that between ionic and polymeric, is somewhat arbitrary.

In group 13, all structures can be described as polymeric. Again, we will not look at the structures in detail although it is worth mentioning that there are a number of different structural types, most of which involve octahedral coordination of the metal atoms and four-coordinate oxygens.

In group 14 the trends are more interesting and informative. Both the common oxides of carbon, CO and CO_2, are molecular covalent species which are gases under ambient conditions (Fig. 5.10), the latter in striking contrast to SiO_2 which is a polymeric, hard and refractory solid; part of the β-crystobalite structure (one of the polymorphs of SiO_2) is shown in Fig. 5.11.

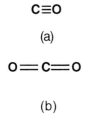

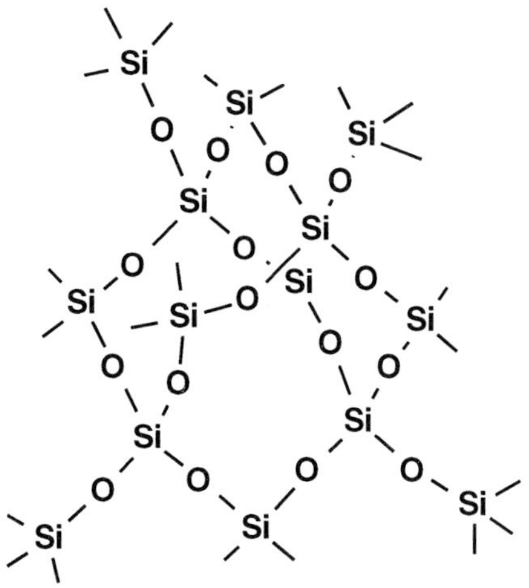

C≡O

(a)

O=C=O

(b)

Fig. 5.10 The structures of (a) CO and (b) CO_2.

Fig. 5.11 Part of the structure of the β-crystobalite form of SiO_2.

The structural differences between CO_2 and SiO_2 are consistent with the respective electronegativity differences between C/O and Si/O, the greater difference being between silicon and oxygen, but this represents only part of the explanation. CO_2 is a linear molecule containing two-coordinate carbon in which strong C–O π-bonding plays an important rôle. On the basis of the discussion in Section 4.3.3 on multiple bonds, this is not surprising, but we would expect that whereas for carbon, a C=O double bond is stronger than two C–O single bonds (2×335 vs. 715 kJ mol^{-1}, Table 4.9), the opposite will be true for silicon oxygen bonds such that silicon will prefer to form two Si–O single bonds rather than one Si=O double bond (2×420 vs. 590 kJ mol^{-1}, Table 4.9). The putative SiO_2 molecule would therefore be expected to polymerize to give a macromolecular solid with four-coordinate silicon and two-coordinate oxygens in which all Si–O bonds are single; this is exactly what is found in the majority of the polymorphs of SiO_2. We should also consider the relative sizes of the elements as well; thus the larger silicon atom is better able to accommodate the larger coordination number, i.e. four vs. two (six-coordinate silicon is also known).

Germanium dioxide has a structural chemistry that is very similar to that of silicon dioxide which is not surprising in view of the similar size and electronegativity of the two elements. In contrast, the structures of tin and lead (one polymorph) dioxide are of the rutile type in which the metal atoms are octahedrally coordinated by oxygen atoms. Other polymorphs of PbO_2 also feature octahedrally coordinated lead and in all cases this increased coordination number of Sn and Pb vs. Ge and Si can be traced to the larger size of the atoms.

The structure of tin monoxide (Fig. 5.12) consists of a planar sheet of oxygens with squares of oxygens alternately capped by tin atoms; the structure of lead monoxide is isomorphous. The lower coordination number of the tin atom (vs. that in SnO_2) results from the presence of the lone pair of electrons associated with the tin(II) centre (c.f. the halides) which presumably resides at the apex of the square pyramid defined by the tin and four oxygen atoms. This structure is sometimes classified as ionic, but the layer or two-dimensional aspect is better described as polymeric since ionic structures are generally more three-dimensional in nature.

The structures of the oxides of the group 15 elements follow a rather similar pattern. Thus all the oxides of nitrogen are molecular covalent, feature strong N–O π-bonding and have coordination numbers around the nitrogen atoms which do not exceed three. The structures are shown in Fig. 5.13.

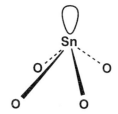

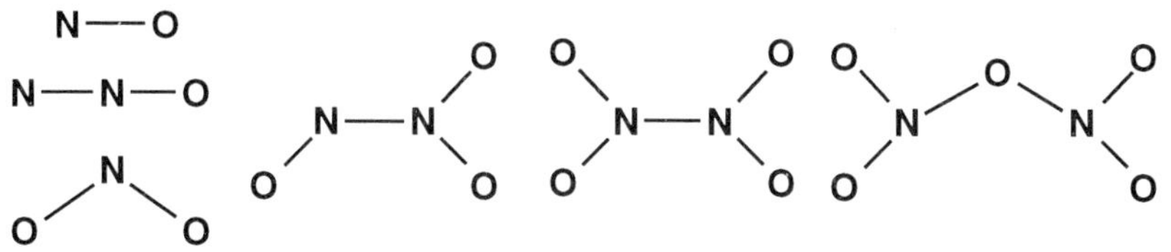

Fig. 5.12 Part of the solid state structure of SnO showing the coordination around the tin atom.

Fig. 5.13 Structures of the oxides of nitrogen; no attempt is made to indicate bond order.

In contrast, none of the molecular nitrogen oxides have any counterpart in phosphorus chemistry (except as high temperature species). The combination of increased electronegativity difference, much weaker P=O π-bonding and the larger size of the phosphorus atom all favour polymeric forms exactly as seen in group 14. The common oxides of phosphorus are P_4O_6 and P_4O_{10}, the structures of which are shown in Fig. 5.14. The term oligomeric might be more appropriate although only one phase of P_4O_{10} contains discrete molecules; two others are also known which comprise two- and three-dimensional macromolecular structures although still based on PO_4 tetrahedra sharing three vertices.

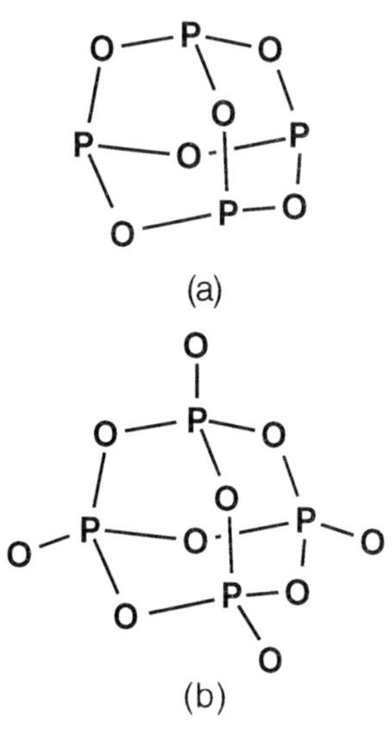

The oxides E_2O_3 of arsenic and antimony both have phases containing molecular E_4O_6 units analogous to P_4O_6 but also exist in polymeric forms consisting of edge-sharing EO_3 pyramids. Bi_2O_3 in contrast, is exclusively polymeric in which the larger bismuth atom attains coordination numbers of

Fig. 5.14 Structures of the oxides of phosphorus, (a) P_4O_6 and (b) P_4O_{10}.

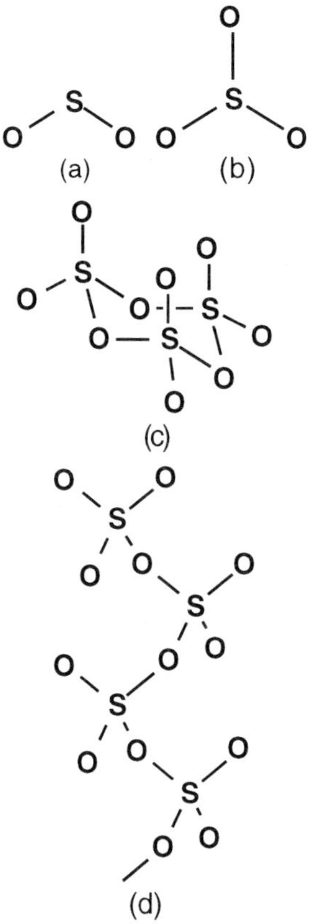

Fig. 5.15 The structures of (a) SO_2, (b) monomeric SO_3, (c) cyclotrimeric γ-SO_3 and (d) helical β-SO_3.

We should note that GeF_2 and SO_2 are isoelectronic, with respect to their valence electrons, and that the bond polarity argument is probably quite appropriate in this case. Germanium is also larger than sulfur but this is unlikely to be important here in view of the ease with which sulfur becomes four-coordinate (as with β- and γ-SO_3). Note that whilst conventional electron counting procedures would assign GeF_2 a vacant orbital, this would not be the case with SO_2.

five in a variety of polymorphs. The pentavalent oxides of arsenic, antimony and bismuth are less well characterized but are certainly polymeric and contain elements with octahedral coordination.

The oxides of the group 16 elements provide another interesting series which illustrates the effects of element size, electronegativity differences and multiple bonding. The elemental forms of oxygen (i.e. the 'oxides' of oxygen) are O_2 and O_3 for both of which multiple bonding is important with concomitant low coordination numbers. With sulfur, the similar electronegativity to oxygen favours the formation of molecular covalent species but the relatively small size of the sulfur atom means that π-bonding is still fairly strong since orbital overlap is good. As a consequence, the most important oxide of sulfur, SO_2, is not only molecular covalent but also contains S–O π-bonding. The other main oxide is SO_3 which is molecular in the gas phase, but exists as a cyclotrimer in the solid γ-phase or as a helical polymer in the β-phase; these structures are shown in Fig. 5.15.

It is worth commenting on why SO_2 is exclusively monomeric whereas SO_3, in the solid state, is polymeric since this would appear to be at odds with what we observed for the germanium fluorides, GeF_2 and GeF_4, i.e. that the lower oxidation states are the ones which tend to polymerize. For SO_3, a possible explanation is that the presence of three electron-withdrawing oxygens around the sulfur in SO_3 induces a sufficiently large δ^+ charge at the sulfur centre to promote oligomerization and that the coordination number is readily increased from three to four. In effect, the sulfur in SO_3 is coordinatively unsaturated whereas the germanium centre in GeF_4 is not. In this regard, we may note that trigonal planar, three-coordination is unusual for sulfur. The reason why GeF_2 polymerizes whereas SO_2 does not is probably a consequence of the much more polar bonds in the former. We may note also that SiO_3^{2-} and PO_3^-, which are isoelectronic (having the same number of electrons) with SO_3, are exclusively polymeric in the solid state in contrast to their lighter congeners CO_3^{2-} and NO_3^- which are always monomeric; size effects are important here together with the stability of multiple bonding when first row elements are involved.

Selenium dioxide, SeO_2, in contrast to SO_2 is polymeric with a linear zig-zag structure shown in Fig. 5.16. We should note that weaker π-bonding and the larger size of the selenium atom are the likely reasons for this difference; the electronegativity difference between the elements is very small.

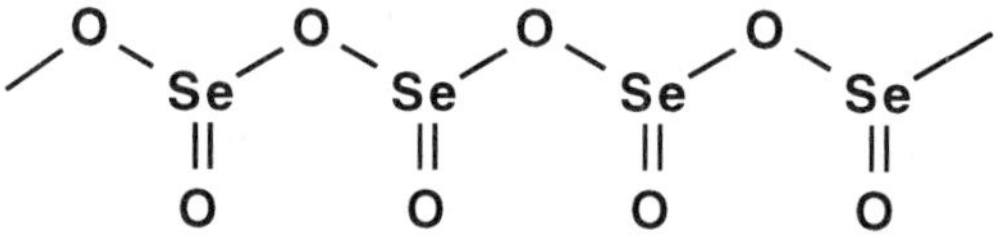

Fig. 5.16 Part of the structure of SeO_2.

Tellurium dioxide is also polymeric although the coordination number of the tellurium atoms in the various polymorphs which are known, is four, and all Te–O bonds are single. Polonium dioxide has the fluorite structure in which the Po atoms are eight-coordinate and the structure can be described as ionic. Note that the dioxides of group 16 clearly illustrate the trend to larger coordination numbers as the group is descended; two for SO_2, three for SeO_2, four for TeO_2 and eight for PoO_2, and the trend from covalent molecules through polymeric structures to an ionic compound.

The higher oxidation state oxides are SeO_3 and TeO_3 and both are polymeric; the former is a cyclotetramer (in contrast to cyclotrimeric SO_3) whereas the latter consists of TeO_6 octahedra sharing all vertices to give a three-dimensional lattice.

In group 17, the oxides of fluorine, chlorine and bromine are all molecular covalent, and π-bonding is important to some extent. The oxides of iodine are all oligomeric or polymeric, consistent with the larger size of the element and its lower electronegativity. Selected structures are shown in Fig. 5.17.

In group 18 the only stable neutral oxides are those of xenon, XeO_3 and XeO_4, which are both molecular.

5.3 Element hydrides

A structural classification of the s- and p-block hydrides is shown in Table 5.4 and the general patterns which we have seen for the halides and oxides are also apparent here. With the exception of BeH_2, the hydrides of groups 1 and 2 all adopt typically ionic structures and also have lattice energies broadly consistent with predominantly ionic bonding. The exceptions to this are the hydrides of the lighter elements such as LiH and MgH_2 which have some degree of covalency, although this is to be expected in view of the high polarizing power of these small cations and the high polarizability of the hydride anion.

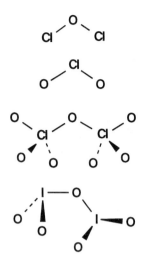

Fig. 5.17 Some oxides of chlorine and iodine.

Table 5.4 A structural classification of the s- and p-block element hydrides

1	2	13	14	15	16	17
LiH	BeH_2	BH_3	CH_4	NH_3	H_2O	HF
NaH	MgH_2	AlH_3	SiH_4	PH_3	H_2S	HCl
KH	CaH_2	GaH_3	GeH_4	AsH_3	H_2Se	HBr
RbH	SrH_2		SnH_4	SbH_3	H_2Te	HI
CsH	BaH_2		PbH_4	BiH_3		

Ionic	Polymeric	Molecular covalent

Beryllium hydride, BeH_2, is best described as polymeric and adopts a linear structure in the solid state with bridging hydrides (similar to $BeCl_2$), the extreme polarizing ability of the hypothetical Be^{2+} ion would make the ionic model inappropriate here in contrast to the hydrides of the heavier group 2 metals.

In group 13, the hydrides are also polymeric although we must be a little careful here in the use of electronegativity arguments (c.f. AlF_3 earlier). Boron forms a very large array of binary hydrides which can only be satisfactorily described in terms of delocalized, multicentre bonding; we shall just look at the simplest hydride, BH_3. Borane is a gas under ambient conditions and exists as a dimer diborane, B_2H_6, which has the hydride-bridged structure shown in Fig. 5.18.

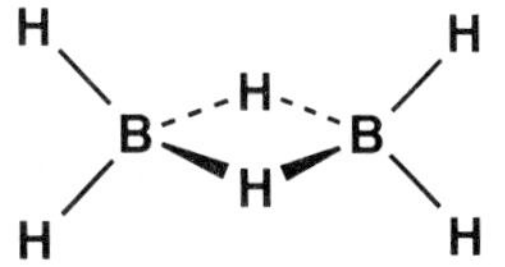

The ionic radius of the hydride anion, H^-, is rather variable depending on its environment but is usually placed between bromide and iodide. Thus it is quite large and this reflects the relative ineffectiveness of a single proton to hold onto two electrons and the resulting dominance of interelectron repulsions. It is therefore not surprising that the hydride anion is easily polarized.

Fig. 5.18 The structure of B_2H_6.

We can stretch our use of the word polymer to a dimer if we wish (we did for I_2Cl_6), but it is important to realize that the dimerization is mostly a consequence of the presence of a vacant orbital on the boron centre rather than any difference in electronegativity between the elements. In fact boron and hydrogen have very similar electronegativities (2.04 and 2.02 respectively on the Pauling scale) and therefore relatively non-polar B–H bonds. A similar situation is found for gallane which is isostructural with diborane, and for similar reasons; the Pauling electronegativity of gallium is 1.81.

Binary, neutral hydrides of indium and thallium are not well characterized as a result of the extreme weakness of the covalent In–H and Tl–H bonds; a factor in common with the hydrides of all the heavier p-block elements. Aluminium hydride is a volatile solid under ambient conditions which can certainly be described as polymeric since it exists as a three-dimensional, hydrogen-bridged lattice. This undoubtedly does reflect the electronegativity difference between the elements which is large enough for a polymeric structure to be expected, although the vacant aluminium 3p orbitals are also important.

The hydrides of group 14, and indeed of all subsequent groups, are all molecular covalent. Methane, CH_4, is the simplest hydride of carbon but there are an essentially limitless number of higher hydrocarbons in which catenation or carbon–carbon bonds are present. Silane, SiH_4, and germane, GeH_4, are the prototypical hydrides of silicon and germanium although the higher catenated congeners are generally less stable than for carbon, this being particularly the case for germanium. The tin and lead hydrides, SnH_4 and PbH_4, are considerably less stable, which again reflects the weaker heavier-element-to-hydrogen bonds; catenated compounds are particularly unstable, being unknown for lead. It is also worth noting that tin(II) and lead(II) hydrides are unknown, particularly in view of the importance of this oxidation state in the chemistry of these elements, especially lead. This is analogous to the situation noted earlier for the lead alkyls. Part of the reason for the instability of the heavier p-block hydrides is the poor overlap between the orbitals of these elements and the 1s orbital of hydrogen. For tin(IV) and lead(IV), there is sufficient overlap for marginal stability but this is not the case for the larger tin(II) and lead(II) with bigger and more diffuse orbitals.

In group 15, the trihydride, EH_3, is known for all elements with stability decreasing down the group for reasons already discussed; only for nitrogen and phosphorus are higher hydrides involving catenation known such as N_2H_4 (hydrazine) and P_2H_4. Conspicuous by their absence are any stable hydrides of the higher +5 oxidation state which is probably partly due to the fact that hydride is fairly readily oxidized and is therefore not compatible with elements in high formal oxidation states. Another important additional factor, however, is that the weaker element–hydrogen bonds which would be expected for a pentavalent hydride would probably result in the reaction shown in Eqn 5.1 being quite exothermic being driven, in part, by the strong H–H bond.

$$EH_5 \quad \rightarrow \quad EH_3 \quad + \quad H_2 \tag{5.1}$$

AlH_3 has some properties in common with some of the s-block hydrides such that an ionic description is sometimes prefered. Indeed, the α-form is isostructural with AlF_3.

Downs and Pulham (*Chem.Soc.Rev.*, (1994), 175) have argued that whilst the weakness of the E–H bonds of the heavier p-block hydrides is an important factor regarding their stability, it is not the whole story since many of the s-block hydrides actually have weaker E–H bonds than some of the heavier hydrides of groups 13 and 14. Also important is the strength of the bridging E–H–E interactions. These are strong for the s-block hydrides since they are essentially ionic in nature, but are much weaker in group 13, for example, where they are more covalent and decrease down the group.

As with the alkyls, the disproportionation reaction $2\,PbH_2 \rightarrow Pb + PbH_4$ is calculated to be exothermic.

The group 16 hydrides are EH_2 (now usually written as H_2E) and again stability decreases down the group; H_2Po is not stable under ambient conditions. The only stable catenated hydride in this group is hydrogen peroxide, H_2O_2, and hydrides of the higher +4 and +6 oxidation states are unknown as in group 15.

Group 17 is characterized by the hydrogen halides, HE (usually HX), with no higher oxidation state hydrides, and there are no stable hydrides of the group 18 elements.

We will finish this section with a brief discussion of the physical nature of the element hydrides of groups 14 and 16. As shown in Table 5.4, they are all molecular covalent species but the trends in melting and boiling points are worth commenting upon. In group 14, we see a progression to higher melting and boiling points as the group is descended resulting from the increasing van der Waals forces between the molecules. This is also largely the case for the group 16 hydrides with the obvious exception of H_2O. The explanation for this feature is the extensive hydrogen bonding between oxygen lones pairs and hydrogen, this being that much greater for the lighter element due to its high electronegativity. Values for all compounds are shown in Fig. 5.19. The values for the group 15 elements show a similar trend to those of group 16 as a result of the importance of hydrogen bonding in NH_3.

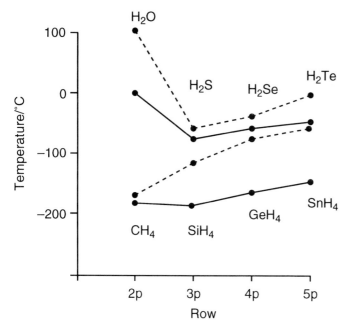

Fig. 5.19 Boiling points (dashed lines) and melting points(unbroken lines) for the hydrides of the group 14 elements (lower two lines) and the group 16 elements (upper two lines).

6 Acids and bases

6.1 Definitions

There are many definitions of acids and bases, some more specific than others although this is not the place to cover the topic in great detail. However, some simple definitions of acidity and basicity will be useful at this point before moving on to look at the various trends which occur primarily in the p-block.

6.1.1 Lewis acidity

The broadest definition of acids and bases is the *Lewis* definition. In this scheme a Lewis acid is an electron-pair acceptor whilst a Lewis base is an electron-pair donor. Two examples will illustrate this model. In the adduct formed between boron trifluoride, BF_3, and trimethylamine, NMe_3, the boron centre acts as a Lewis acid whilst the nitrogen centre acts as a Lewis base. The resulting N–B bond is called a coordinate or dative covalent bond and is represented by an arrow as shown in Fig. 6.1. The nitrogen is able to act as a base since it has a non-bonded or lone pair of electrons and the boron centre is acidic because it posseses a vacant p orbital which can accept an electron pair.

As a second example we can look at the interaction between SiF_4 and pyridine which results in the 2:1 adduct, $[SiF_4(py)_2]$ (Fig. 6.2). The nature of the vacant orbital on the silicon centre is something to which we shall return in section 6.3.

Fig. 6.1 The structure of $Me_3N \rightarrow BF_3$.

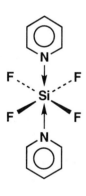

Fig. 6.2 The structure of $SiF_4(pyridine)_2$.

Note that in aqueous solution a proton does not exist in isolation; it is certainly hydrated and is generally written as H_3O^+ although this is also an oversimplification. Larger aggregates such as $H_5O_2^+$ are undoubtedly present as well.

6.1.2 Brønsted–Lowry acidity

The *Brønsted–Lowry* concept is probably the second most commonly employed model for acids and bases although in aqueous solution it is by far the most widely used. In this scheme an acid is defined as a proton (H^+) donor and a base as a proton acceptor. We can illustrate this with reference to the familiar phrase 'acid plus base gives salt plus water' an example of which is given in Eqn 6.1.

$$HCl\ (aq)\ +\ NaOH\ (aq)\ \rightarrow\ NaCl\ (aq)\ +\ H_2O\ (l) \tag{6.1}$$

Here hydrochloric acid, HCl, is the acid, sodium hydroxide the base and sodium chloride the salt or neutralization product.

This system can readily be extended to other protonic solvent systems such as sulfuric acid and liquid ammonia. Moreover, the concept of conjugate acids and conjugate bases is also useful as indicated in Eqn 6.2.

$$HS^-\ (aq)\ +\ H_2O\ (l)\ \rightarrow\ S^{2-}\ (aq)\ +\ H_3O^+\ (aq) \tag{6.2}$$

Thus S^{2-} is the conjugate base of the acid HS^- and H_3O^+ is the conjugate acid of the base H_2O. Note that HS^- can act as an acid or as a base depending on the conditions or pH. Under acidic conditions it acts as a base and is protonated to give H_2S whilst under basic conditions it acts as an acid and is deprotonated to give S^{2-}.

6.1.3 Lux–Flood acidity

A final definition which will be useful to us is the *Lux–Flood* model in which an acid is defined as an oxide acceptor and a base as an oxide donor. This is illustrated by the following example in Eqn 6.3 where CaO is the base, SiO_2 is the acid and calcium silicate is the neutralization product.

$$CaO \ + \ SiO_2 \ \rightarrow \ CaSiO_3 \qquad (6.3)$$

Note that the concept of Lewis acidity and basicity incorporates both the Brønsted–Lowry and Lux–Flood approaches as special cases.

6.2 Element oxides and hydroxides

One important aspect of the element oxides, discussed in Chapter 5, and of the closely related hydroxides, concerns their acidic or basic properties and the relationship between these properties and the position of the element in the periodic table. Figure 6.3 illustrates these ideas and we will consider it in detail shortly, but first we should understand what is meant by an oxide (or hydroxide) being acidic or basic and also what is meant by the term *amphoteric* which is often used in association with these compounds.

An acidic oxide, in aqueous solution, is one which combines with a water molecule and then releases a proton with concomitant formation of an oxo-anion. This is illustrated in Eqn 6.4 for sulphur trioxide, SO_3:

$$SO_3 \text{ (aq)} \ + \ H_2O \text{ (l)} \ \rightarrow \ H_2SO_4 \text{ (aq)} \ \rightarrow \ H^+ \text{ (aq)} \ + \ HSO_4^- \text{ (aq)}$$
$$\rightarrow \ 2H^+ \text{ (aq)} \ + \ SO_4^{2-} \text{ (aq)} \qquad (6.4)$$

Thus addition of SO_3 to water forms sulphuric acid, H_2SO_4, which can lose one proton to form bisulphate, HSO_4^-, or two protons to form sulphate, SO_4^{2-}.

All of the elements in bold face in Figure 6.3 have acidic oxides and therefore react with water to give acidic solutions (because they contain H^+) and oxo-element anions. Clearly an element hydroxide is also considered acidic if it dissociates in water to give H^+ and an oxo-anion, although in some cases, hydroxides are acidic not as a result of being proton donors, but rather because they act as hydroxide acceptors. An example of this is found with boron trihydroxide or orthoboric acid, $B(OH)_3$ or H_3BO_3. The former formula is more appropriate since $B(OH)_3$ is not a source of protons in aqueous solution, but acts instead as a hydroxide acceptor affording the $[B(OH)_4]^-$ anion.

An alternative description is that an acidic oxide reacts with an aqueous base, i.e. it acts as a hydroxide acceptor as discussed above, to give an

Note that SO_3 is also acidic according to the Lewis and Lux–Flood criteria as exemplified by the equation below which shows the addition of oxide to form sulphate:
$$SO_3 \ + \ O^{2-} \ \rightarrow \ SO_4^{2-}$$
As an oxide acceptor, it can be classified as a Lux–Flood acid whereas in accepting an electron pair from oxide, it also acts as an acid according to the Lewis definition.

oxo-anion (or hydroxo-anion); using the same system as above, we can represent this as shown in Eqn 6.5.

$$SO_3\ (aq)\ +\ OH^-\ (aq)\ \rightarrow\ HSO_4^-\ (aq)\ \rightarrow\ H^+\ (aq)\ +\ SO_4^{2-}\ (aq) \quad (6.5)$$

A basic oxide reacts with water and dissociates to give an element cation and hydroxide ions as illustrated for calcium oxide in Eqn 6.6.

$$CaO\ (s)\ +\ H_2O\ (l)\ \rightarrow\ Ca^{2+}\ (aq)\ +\ 2OH^-\ (aq) \quad (6.6)$$

An element hydroxide similarly dissociates in water to give an element cation and hydroxide ions although we should note that the distinction between oxides and hydroxides in aqueous solution is often more formal than real. Thus addition of H_2O to CaO will initially afford $Ca(OH)_2$ which then subsequently dissociates. The elements at the lower left in Figure 6.3 (in normal face) all form basic oxides and hydroxides.

As with the acids above, we can employ an alternative description wherein a basic oxide reacts with aqueous acid to form an element cation and water as shown in Eqn 6.7.

$$CaO\ (s)\ +\ 2H^+\ (aq)\ \rightarrow\ Ca^{2+}\ (aq)\ +\ H_2O\ (l) \quad (6.7)$$

While on the subject of acids, an interesting comparison can be made between the dihydrates of H_2SO_4 and H_2TeO_4. In the former case, this is simply a hydrogen bonded adduct of molecular H_2SO_4 whereas in the latter case, H_2O adds across the formal Te=O bonds to give H_6TeO_6 (or $Te(OH)_6$) which is an octahedral hexahydroxide of Te. The difference between S and Te can be traced to three factors: (i) the larger Te can more readily support a larger coordination number, (ii) Te prefers two Te–O single bonds to one Te=O double bond in contrast to S and (iii) the initial Te=O bond is more polar than the S=O bond thereby enhancing the addition of H_2O.

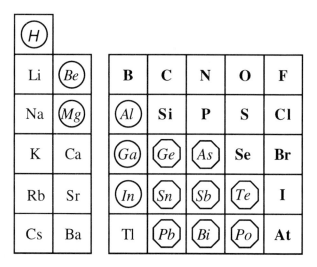

Fig. 6.3 The s- and p-block elements indicating the acidity, basicity or amphoterism of their oxides: elements in bold face are acidic, in normal face are basic, in circles are amphoteric in all oxidation states and in octagons are amphoteric in their lower oxidation states and acidic in their higher oxidation states.

An amphoteric oxide or hydroxide can react with both acids and bases and thus its acidity or basicity depends on whether it is in acid or alkaline solution. We can illustrate this by considering the reactions of aluminium oxide, Al_2O_3, shown in Eqns 6.8 and 6.9.

$$(basic)\quad Al_2O_3\ (s)\ +\ 6H^+\ (aq)\ \rightarrow\ 2Al^{3+}\ (aq)\ +\ 3H_2O\ (l) \quad (6.8)$$

(acidic) Al_2O_3 (s) + $2OH^-$ (aq) + $3H_2O$ → $2[Al(OH)_4]^-$ (aq)   (6.9)

In Eqn 6.8, aluminium oxide acts as a base under acidic conditions with the formation of the element cation, Al^{3+}. Under basic conditions as shown in Eqn 6.9, it behaves as an acid (note that it acts as a hydroxide acceptor rather than a proton donor in this case) with the formation of an element hydroxo-anion.

In Figure 6.3, the oxides of the elements in circles are amphoteric in all of their oxidation states whilst those in octagons have acidic oxides in their maximum oxidation states and amphoteric oxides in their lower oxidation states.

It is clear from Figure 6.3 that elements in the s-block (with the exception of hydrogen) and lower left of the p-block have basic oxides and hydroxides, those in the top right of the p-block are acidic and elements along an approximate diagonal from top left to bottom right tend to be amphoteric. Moreover, there is an obvious parallel between this trend and that observed for the electronegativity of the elements, and it is on this basis that we can seek an explanation. Thus electronegative elements form acidic oxides, electropositive elements form basic oxides and amphoterism is observed for elements of intermediate electronegativity. We can understand why this should be the case by considering an element hydroxide unit, E–O–H.

If E is very electronegative, the electron density will tend to be polarized towards the E centre, which we can represent as δ^- E–O–H δ^+, or certainly towards the E–O unit, δ^- (E–O)–H δ^+. This polarization facilitates loss of a proton (H^+), or O–H bond heterolysis, and the negative charge of the resulting oxo-anion is effectively stabilized by the presence of the electronegative element E, i.e. it is not strongly localized at the oxygen atom. In addition, an electronegative element is unlikely to form a cation, E^+. In contrast, if E has a low electronegativity, i.e. if it is an electropositive element, the polarization will tend to be in the opposite sense, i.e. δ^+ E–O(δ^-)–H or δ^+ E–(O–H) δ^-, such that dissociation occurs by heterolysis of the E–O bond to give E^+ and OH^-. The formation of cations in aqueous solution is a common feature of electropositive, metallic elements.

In elements of intermediate electronegativity, the magnitude of any bond polarization is insufficient to result in the predominance of either acidic or basic behaviour and amphoteric properties are observed. However, this is often dependent on the element oxidation state. The elements Be, Mg, Al, Ga and In are amphoteric in all of their oxidation states although this is a somewhat misleading statement since only for the group 13 elements is there a possibility of more than one oxidation state. Moreover, stable oxides and hydroxides for the three group 3 elements listed are only found for the +3 state. With the elements Ge, Sn, Pb, As, Sb and Bi, the oxides and hydroxides are acidic in their highest oxidation states and amphoteric in lower oxidation states, e.g. arsenic(III) oxide, As_2O_3, is amphoteric whereas arsenic(V) oxide is acidic. We can account for this observation if we recall that element electronegativity is, to some extent, a function of oxidation

A interesting point relating to the properties of the isoelectronic oxo-anions ClO_4^-, SO_4^{2-}, PO_4^{3-} and SiO_4^{4-} is that there is an increasing tendency towards condensation in the presence of acid (H^+) to give di- or oligonuclear oxo-bridged species as this relieves the increasing build-up of charge.

state and that the higher the oxidation state, the higher the electronegativity.
Note that hydrogen forms an amphoteric oxide, i.e. H_2O.

We may also consider the trends in the strengths of various acids in aqueous solution as indicated by their pK_a values. Table 6.1 shows the pK_a values for the four oxo-acids of chlorine from which it is clear that the acid strength increases dramatically as the number of oxygen atoms attached to the chlorine atom increases. This is simply a result of more oxygens being able to more effectively delocalize the negative charge (or, in other words, to stabilize and render less basic the conjugate base).

A further point concerns the trend in the acidities of the p-block element hydrides which are shown in Table 6.2.

Table 6.1 pK_a values for the oxo-acids of chlorine

Acid	pK_a
$Cl(OH)$	7.2
$ClO(OH)$	2.0
$ClO_2(OH)$	−1
$ClO_3(OH)$	−10

Table 6.2 pK_a values for some of the p-block element hydrides

14		15		16		17	
CH_4	46	NH_3	35	OH_2	16	FH	3
SiH_4	35	PH_3	27	SH_2	7	ClH	−7
GeH_4	25	AsH_3	23	SeH_2	4	BrH	−9
				TeH_2	3	IH	−10

We note that the acidity increases on moving from left to right and also on descending a group. These observations cannot be rationalized by a single principle such as electronegativity or bond strength, but it is clear that for a given row, the element–hydrogen electronegativity difference is the dominant factor whereas in a group, it is the element–hydrogen bond strength which is the most important. In the latter case, weaker bonds result in more ready dissociation and these values show much the same trend as those measured for the gas phase acidities. In the former case, however, the gas phase acidities are found to be fairly constant across a row implying that in aqueous solution it is not so much the electronegativity difference which is important but rather the increasing hydration energy associated with smaller, more highly charged anions which have a greater charge to size ratio or charge density.

6.3 Lewis acidity of the heavier p-block elements

6.3.1 General points

In Section 6.1.1 the examples used to illustrate Lewis acidity were BF_3 and SiF_4 and it will be useful to consider the origin of this acidity in both compounds. In the former, the three-coordinate, trivalent boron centre has six electrons (three from itself, one each from the fluorines) but with four valence orbitals (one s and three p) a total of eight can be accommodated. Thus, boron in BF_3 has a vacant valence orbital which can readily accept an electron pair, through formation of a coordinate or dative covalent bond, resulting in a species with eight valence electrons. In fact, we should remember for BF_3 that there will be some interaction between the vacant boron 2p orbital and filled π-type orbitals on the fluorine atoms, i.e. some

B–X (X = halide) π-bonding should be greater, the smaller the halide (due to a more favourable orbital size match) which is the main reason usually offered to account for the obsevation that, for the boron trihalides BX_3, Lewis acidity increases in the order $BF_3 < BCl_3 < BBr_3 < BI_3$; this is the opposite trend to that which would be expected on the basis of element electronegativity differences.
Note that the isoelectronic BO_3^{3-} is also a Lewis acid but that CO_3^{2-} and NO_3^-, which are also isoelectronic, are not acidic. This reflects the lower difference in electronegativity between C/N and O and also the stronger C–O and N–O π-bonding.

B–F π-bonding will be present (which can also be considered as an acid–base interaction where the boron is described as a π acid and fluorine as a π base) such that the boron is not really electron deficient in this sense. Nevertheless, the formation of a new σ-bond will readily displace any π-bonding electrons in this case.

In fact, Lewis acidity is a general feature of the chemistry of the group 13 elements, and of the elements of groups 1 and 2, which have fewer valence electrons than orbitals. Moreover, we can think of compounds with eight electrons as electronically saturated since all four valence orbitals are associated with 2-centre-2-electron bonds (this idea is, of course, the basis of the Lewis octet rule). Compounds with fewer than eight electrons can therefore be thought of as unsaturated or electron deficient with a resulting tendency to form extra bonds to electron donors.

If we now turn to the example of SiF_4, it may not at first be clear why this molecule should act as a Lewis acid since it is already an 8-electron species, and therefore electronically saturated according to the definition given above. Nevertheless, SiF_4 forms an adduct with pyridine (Fig. 6.2) and also with fluoride to give a complex anion of the formula SiF_6^{2-} both of which can be considered as 12-electron species.

In seeking an explanation for the acidity of SiF_4 we should first consider that the carbon analogue, CF_4, shows no such acidic behaviour and, what is more, it is a general feature of the chemistry of the first row elements that the octet rule is not violated. Moreover, the acidity of SiF_4 and the 'apparent' violation of the octet rule is not an isolated example, but is a common feature of the chemistry of the heavier p-block elements. If we take CF_4 and SiF_4 as examples, two factors which we should initially consider are, firstly, that silicon is bigger than carbon and secondly, that the Si–F bonds will be more polar than the C–F bonds. As we have seen before, larger atoms can support larger coordination numbers and a more δ^+ silicon centre is certainly likely to be more acidic.

Whilst these ideas of size and polarity are important, they are not a sufficient explanation for the acidity of SiF_4 or, indeed, for the Lewis acidity of a large number of compounds of the heavier p-block elements. This is because the bonding between silicon and the pyridine ligands in $[SiF_4(py)_2]$ is not ionic or electrostatic; there is a high degree of covalency, and for this to result, there must be some suitable orbitals associated with the silicon centre. In other words, SiF_4 must have some low energy or low-lying vacant orbitals with which it can form additional covalent interactions with Lewis bases. The nature of these orbitals is not unambiguous and we shall address two particular views in turn.

6.3.2 Vacant d orbitals

The traditional view of bonding in a molecule like $[SiF_4(py)_2]$ is to assume that all bonds are of the standard 2-centre-2-electron type and therefore that there are twelve electrons associated with the silicon centre (four Si–F bonds and two dative Si–N bonds). This feature was mentioned above and is why the compounds are said to violate the octet rule. Since silicon only has four valence orbitals, which can accommodate a maximum of eight electrons,

The two factors of size and bond polarity can also account for the related fact that SiF_4 and $SiCl_4$ are readily hydrolysed whereas CF_4 CCl_4 are not. H_2O is a Lewis base and can coordinate to the silicon centre whereas this is not possible in the case of carbon. Moreover, subsequent HF or HCl elimination is facilitated by the close proximity of the H and F or Cl atoms in the silicon coordination complex and also by the more δ^- F or Cl centres.

There is an interesting aspect of Lewis acidity which bears on the matter of the stability of the group maximum oxidation state for the 4p elements (Section 4.1). For example, $AsCl_5$ decomposes above $-50°C$ into $AsCl_3$ and Cl_2, but the acid-base complex $AsCl_5(OPMe_3)$, wherein the Lewis acidic $AsCl_5$ is complexed by the base trimethylphosphine oxide, is stable up to $+50°C$. There are, in fact, many examples where an apparently unstable oxidation state can be stabilised by complexation with Lewis bases which suggests that the normally facile decomposition modes are kinetic in origin and can be arrested by effectively saturating the coordination sphere of the central element. A further, more general, example is that it is often observed that for oxidation states where a neutral fluoride or oxide is unstable, the related oxo or fluoro-anions are often much more stable. Part of the reason given for this, is that the negative charge raises the energy of the O or F σ orbitals which leads to better overlap with the central element centre and stronger bonds result. Coordination number arguments and kinetic effects will also be important.

other orbitals must be involved and it is assumed that these are vacant, low-lying 3d orbitals. It is then be possible to use two 3d orbitals (usually d_{z^2} and $d_{x^2-y^2}$) as well as the 3s and three 3p orbitals to construct a set of six d^2sp^3 hybrids with which to form six 2-centre-2-electron bonds. Moreover, the hybrids chosen are equivalent and point to the vertices of an octahedron which accounts for both the shape and the six equal Si–F bond lengths in SiF_6^{2-}. In employing vacant d orbitals, the problems with violation of the octet rule (hypervalence as it is often called) in compounds of the heavier p-block elements compounds are circumvented, yet the lack of any such compounds in the chemistry of the 2p elements is readily understood in view of the absence of any 2d orbitals.

This model employing d orbitals is quite useful and has achieved considerable currency in inorganic chemical thinking. Furthermore, we can develop the ideas to account for the different bond lengths found in a trigonal bipyramidal molecule such as PF_5, an example of a hypervalent, 10-electron phosphorus compound. In PF_5, 2-centre-2-electron bonding requires the use of one d orbital to construct a set of dsp^3 hybrids, although it is not possible to generate five equivalent hybrids. In fact, the hybrids are constructed in two sets; an sp^2 set which form bonds to the three equatorial fluorines, and a dp set which are involved in bonding to the axial atoms. Since electrons are most tightly held in s orbitals (since they are lower in energy), hybrid orbitals with an s component tend to be smaller (less diffuse) than those without, with the consequence that bonds to orbitals which contain any degree of s character tend to be shorter (the same is true for p vs. d orbitals). This model accounts in a simple way for the observed shorter equatorial P–F bonds in PF_5 although it should be noted that shorter equatorial bonds tends to be a feature of compounds with smaller central element centres; in trigonal bipyramidal $BiMe_5$, all Bi–C bonds are equal in length.

A final molecule which we shall consider in this section is the triiodide anion, I_3^-. This ion is readily formed from diiodine, I_2, and iodide, I^-, which can be viewed as an acid-base reaction where I_2 is the Lewis acid. We can assume that I_2 is a Lewis acid due to the presence of vacant d orbitals and further, that the hybridization of the resulting central iodine in the linear I_3^- is dsp^3 with bonds formed to the two terminal iodines using the axial dp set. The three lone pairs then reside in the equatorial sp^2 orbitals. The longer observed I–I bonds in I_3^- compared to that in I_2 can also be accounted for since, in the former they involve dp hybrids whereas in the latter, the I–I bond is formed from sp hybrids (or at least hybrids with some s character).

In considering I_3^-, we described the structure as linear; this is what is observed and it is predicted on the basis of valence shell electron pair repulsion (VSEPR) theory which has been extensively developed by Gillespie and which we will briefly discuss in the following chapter. Indeed, VSEPR is an extremely powerful model for rationalizing the structures of molecules in the p-block. It explicitly employs the concept of 2-centre-2-electron bonds, or more generally, electron-pair domains (lone pairs are considered as well), and thus, implicitly, the concept of d orbital hybridization in hypervalent compounds. Herein, however, lies one of the problems with the d orbital model for Lewis acidity. The structures of

The deconstruction of dsp^3 hybrids into an sp^2 set and a dp set also accounts for the axial site preference in trigonal bipyramidal molecules; electronegative atoms or groups will tend to bond to hybrids in which the electrons are less tightly held. This is an example of Bent's rule which is explained later. A fuller discussion of many aspects of hybrid orbitals is given by Winter (1993).

It should be stressed that hybridisation need only be introduced after the fact in discussions of VSEPR, it is not really required. However, if we wish to equate electron-pair domains with particular hybrids, d orbitals are needed for compounds with more than four electron pairs.

hypervalent molecules can be accounted for on the basis of VSEPR and a certain hybridization can then be assumed after the fact, i.e. dsp^3, d^2sp^3, etc., but we cannot easily use the d orbital model of itself to predict structure. The central iodine in I_3^- can be described as dsp^3 hybridized but the use of one particular d orbital and the deconstruction to sp^2 and dp sets is rather arbitrary. It is the same as assuming (wrongly) that methane is tetrahedral *because* it is sp^3 hybridized; in fact methane is tetrahedral because this is the lowest energy configuration for the molecule, and sp^3 hybridization can then be used (as one model) to describe the bonding.

In other words, we cannot use d orbitals in themselves to account for the structures of molecules such as I_3^- but rather must use them in conjunction with VSEPR. Whilst this is not a serious problem, a second problem makes the d orbital argument much less tenable since it is now generally accepted that d orbitals are too high in energy to be used effectively in bonding in the p-block. An alternative explanation for the Lewis acidity of the heavier p-block elements is therefore desirable and as we shall see in the following section, there is a model which accounts for Lewis acidity, structure and bond lengths.

6.3.3 Vacant σ* orbitals

Let us consider I_3^- again. We can, in fact, construct a bonding or molecular orbital picture for this molecule in which d orbitals are unnecessary. If we assume that the central iodine has three lone pairs in three sp^2 orbitals, there is one remaining p orbital which can be used to form a multicentre bond to suitably hybridized σ orbitals of the two terminal iodines as represented in Fig. 6.4.

A growing number of high level *ab initio* theoretical studies indicate that vacant d orbitals are too high in energy to play an important role in bonding. This is reflected in the calculated very low occupancy of the d orbitals or, more precisely, the very small extent to which d orbitals mix with the valence sp set. Nevertheless, d orbitals *are* important as polarization functions in the calculations although this is *not* the same thing as saying that their electron occupancy is significant.

A fuller discussion of many aspects of molecular orbital theory is given by Winter (1993).

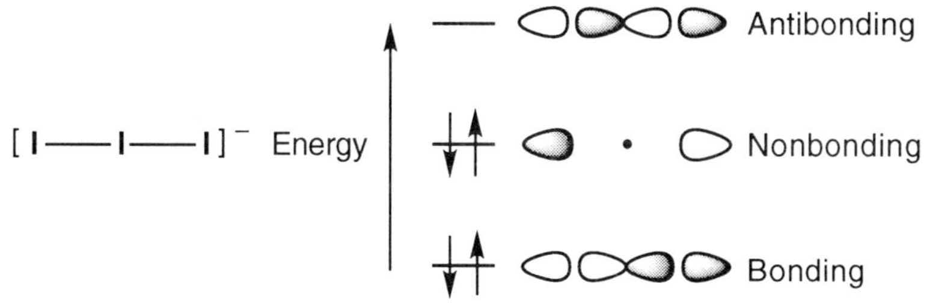

Fig. 6.4 An orbital interaction diagram for I_3^-.

A 3-atom-3-orbital interaction results in which the bonding and non-bonding orbitals are filled and for which the net I–I bond order is one half. The reduced bond order of this 3-centre-4-electron bond thus neatly accounts for the longer I–I bonds without recourse to d orbitals. Moreover, the electron count around the central iodine is effectively eight since there are only three lone pairs and one π-bonding pair associated with the iodine centre. The pair of electrons in the non-bonding π orbital are localized on the terminal iodines and do not contribute to the electron count at the iodine so it can be seen that the octet rule is not really violated at all.

This sort of bonding argument ignoring d orbitals is quite general; it can be used, for example, to account for the longer axial P–F bonds in PF_5 and on this basis, the effective electron count around the phosphorus atom is also eight.

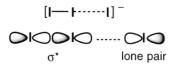

Fig. 6.5 The I–I σ* orbital derived from the overlap of two p_z orbitals.

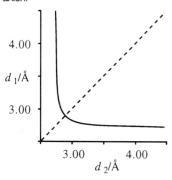

$[I—I\cdots\cdots I]^-$

σ*   lone pair

Fig. 6.6 Orbital interactions in the I_3^- anion.

Fig. 6.7 A graph of the correlation of the I–I·····I bond distances in I_3^- anions; the graph is symmetrical about the dashed diagonal line.

It should also be noted that the curve in Fig. 6.7 can be considered as a reaction coordinate for the nucleophilic displacement of iodide from diiodine by iodide, i.e. I–I + I$^-$ → I$^-$ + I–I.

The structure of BiI$_3$, with symmetrical six-fold coordination around the bismuth centres, is the same as that of AlCl$_3$.

Thus we can abandon d orbitals in describing the bonding in I_3^- but we must still account for the Lewis acidity of I_2. In other words, we must consider the nature of the low-lying vacant orbital(s) of I_2. The most appropriate is the I–I σ* orbital (Fig. 6.5) which is expected to be quite low in energy due to the large and diffuse nature of the iodine orbitals (the corollary is that the bonding orbital will be high in energy which accounts for the relative weakness of the I–I bond).

If we consider the interaction between I_2 and I^-, it is clear that maximum overlap will occur when the iodide approaches *trans* to the I–I bond (Fig. 6.6). Moreover, as the I–I to I^- distance decreases, the population of the I–I σ* orbital will increase with a concomitant lengthening of the I–I bond. In the extreme where both I–I distances are the same, the delocalized bonding scheme described above in Fig. 6.4 is appropriate.

Thus, not only does the σ* model account for the observed geometry of I_3^- and for the longer I–I bonds observed in I_3^- vs. I_2, it also suggests that a correlation should exist between the I to I distances in the unsymmetrical species, I–I·····I, i.e. as the secondary I·····I bond, as it is often called, gets shorter, the primary bond gets longer. This is exactly what is observed in a range of structures involving the I_3^- anion; the I–I·····I distances are correlated and are found to lie on the curve shown in Fig. 6.7.

Thus, the σ* bonding model of Lewis acidity accounts for all of the observations traditionally rationalized on the basis of d orbital participation. Moreover, the structures of Lewis acid–base complexes of the p-block elements are understood in a more direct manner than is the case with d orbitals. This is especially well illustrated in the case of the solid state structures of the tri-iodides of arsenic, antimony and bismuth. All three compounds are isomorphous and can be considered as a close-packed array of iodines with As, Sb or Bi in the octahedral holes. For AsI$_3$ there are three short mutually *cis* As–I bonds with three longer As····I interactions *trans* to these bonds. In SbI$_3$ the basic structure is the same except that the ratio of the primary to the secondary bonds is smaller; for BiI$_3$ the ratio is 1 and all Bi–I bonds are the same length (Fig. 6.8). Note that the *trans* disposition of the secondary bonds to the primary bonds is exactly what would be predicted on the basis of the σ* orbital model.

Moreover, the increasing Lewis acidity of BiI$_3$ vs. AsI$_3$, as exemplified by the more symmetrical structure, i.e. a smaller ratio of primary to secondary bond lengths, is consistent with a lower energy σ* orbital formed by overlap between two large atoms with large diffuse orbitals. We should also note that this model would lead us to expect a regular octahedral structure for the BiI$_6$ unit whereas the VSEPR predicted structure would be ambiguous due to the presence of a lone pair of electrons (see later); each bismuth centre has seven electron pairs and the stereochemical activity or not of the lone pair is always problematical for these types of structure.

We should note that the trend in the structures of AsI$_3$, SbI$_3$ and BiI$_3$ is essentially that from covalent to polymeric but this is slightly counter to that which we would expect on the basis of electronegativity differences (Table 2.5). However, in no case is the element electronegativity difference large, and it is therefore not such a good guide in accounting for this trend.

What is more important here is the trend to increasing Lewis acidity as the group is descended and hence the increasing trend toward strong secondary bonding interactions and polymeric structures. We should note also that the electronegativities of Bi (2.02) and I (2.66) are such as to place the compound near the metalloid region and consistent with this is the fact that BiI_3 is a semiconductor.

In concluding this section, we should note that employing σ^* orbitals readily accounts for the Lewis acidity of the compounds of the heavier main group elements since it is for these elements that orbitals are large and diffuse and for which σ^* orbitals would be expected to be low in energy. For elements of the 2p row, Lewis acidity is not observed because orbital overlap is good and the σ^* orbitals are much higher in energy. Moreover, heavy element iodides would be expected to exhibit the most Lewis acidic behaviour (as seen in the iodides of As, Sb and Bi) since two large elements are involved and bonding occurs between two large and diffuse orbitals.

The low energy of the σ^* orbitals is also important in understanding the observation that iodides are often coloured whereas chlorides and fluorides are not. The $n \rightarrow \sigma^*$ transitions which give rise to the colour tend to occur in the visible region for iodides whereas they are in the ultraviolet for the lighter halides and this also accounts for the photosensitivity of many heavy metal iodides.

Ideas relating to hypervalence and the octet rule are still a matter of much debate. Thus, the neglect of d orbitals in molecular orbital terms does not tend to lead to violations of the octet rule as we have seen. However, if we consider the bonding in, for example, PF_5 or SF_6 to comprise 2-centre, 2-electron P–F or S–F bonds, a total of 10 and 12 valence electrons respectively are required which clearly violate the octet rule and such compounds are often termed hypervalent. Nevertheless, the usefulness of the term hypervalence has been questioned by Gillespie who would rather it be abandoned. In its place would be the octet rule and the duodecet rule for second row molecules like CF_4, and for third and higher row molecules like SF_6 respectively, both of which are chemically stable and have many properties in common. Exceptions would then be the Lewis acidic species like the six-electron BF_3 and the ten-electron PF_5 both of which readily accept fluoride to form the octet and duodecet species $[BF_4]^-$ and $[PF_6]^-$. Quite where this would leave SiF_4, however, is unclear: it has an octet of electrons like CF_4 but can bond to two extra fluorides to give the duodecet compound $[SiF_6]^-$.

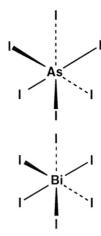

Fig. 6.8 The coordination geometry around arsenic and bismuth in the solid state structures of AsI_3 and BiI_3 respectively.

Note generally that where we have a E–X–E bridging situation, then for a given E, the bridge is more symmetric as X becomes less electronegative and that for a given X, the bridge is more symmetric as E becomes less electronegative. In both cases there is a trend to increasing metallic or metalloid behaviour depending on the absolute elecronegativities of E and X.

6.4 Hard and soft acids and bases

In concluding this chapter, we should say something about the concept of hard and soft acids and bases. In general, both acids and bases can be classified as either hard or soft, and it is found that hard acids preferentially bond to (or form adducts with) hard bases and soft acids to (or with) soft bases. A selection of hard and soft acids and bases are shown in Table 6.3.

Remember that covalent interactions are strongest when orbitals energies are similar.

HOMO means highest occupied molecular orbital and LUMO means lowest unoccupied molecular orbital.

A simple and useful explanation for the attraction of hard to hard and soft to soft is as follows. Hard acids (electron-pair acceptors) have no low lying or low energy unfilled orbitals (i.e. the LUMO is high in energy) and therefore do not readily form covalent bonds with bases. Hard bases (electron-pair donors) have no high-lying or high-energy orbitals (i.e. the HOMO is low in energy) and also do not readily form covalent bonds (to acids), but since the acids and bases are usually positively and negatively charged respectively, they tend to bond strongly to each other through ionic type interactions.

Table 6.3 A selection of hard and soft acids and bases together with some borderline cases

	Acids	
hard	borderline	soft
H^+, Li^+, Be^{2+}, Mg^{2+}, Al^{3+}, Ga^{3+}, In^{3+}, As^{3+}, BF_3, $AlMe_3$, $AlCl_3$, SO_3	Sn^{2+}, Pb^{2+}, Bi^{3+}, BMe_3, SO_2, NO^+, GaH_3	Tl^+, Hg^+, Te^{4+}, Tl^{3+}, $TlMe_3$, BH_3, $GaMe_3$, $GaCl_3$, $InCl_3$, I^+, I_2,
	Bases	
hard	borderline	soft
H_2O, OH^-, F^-, SO_4^{2-}, Cl^-, CO_3^{2-}, ClO_4^-, NO_3^-, NH_3	N_3^-, Br^-, NO_2^-, SO_3^{2-}, N_2	H_2S, I^-, SCN^-, R_3P, CN^-, CO, H^-

In contrast, soft acids do tend to have low-energy unfilled orbitals (a low energy LUMO), which will readily enter in covalent bond formation, and soft bases tend to have high-energy filled orbitals (a high energy HOMO) which will also easily form covalent bonds. The combination of soft acids with soft bases, therefore, is the one which maximizes a covalent interaction.

The general principle of hardness (usually given the symbol η) can be quantified (for atoms, ions or molecules) as shown in Eqn 6.10 where I is the first ionisation energy and A is the first electron affinity.

$$\eta = (I - A)/2 \tag{6.10}$$

In orbital terms, I and A correspond to the energies of the HOMO and LUMO respectively (assuming Koopmans's theorem holds) such that increasing hardness corresponds to an increasing energy separation between these orbitals; the inverse, i.e. softness, $\sigma = 1/\eta$. A similar property, the chemical potential, μ, defined as in Eqn 6.11, is also commonly encountered with respect to this topic, the inverse of which is equal to the Mulliken electronegativity mentioned earlier and is the average of the energies of the HOMO and LUMO in orbital terms.

$$-\mu = (I + A)/2 \tag{6.11}$$

The so-called *principle of maximum hardness* states that chemical systems will tend to be as hard as possible.

7 Structure

7.1 Valence shell electron pair repulsion theory (VSEPR)

The molecules formed by elements of the p-block often appear to present a bewildering array of structures. For example, if we consider some of the trifluorides, BF_3 is trigonal planar, NF_3 is trigonal pyramidal and ClF_3 is T-shaped. Moreover, PF_3 is found to be 'more' pyramidal than NF_3, i.e. the F–P–F angles are smaller than the F–N–F angles. A successful theory of molecular structure must account for all of these observations in a simple and elegant manner, and a theory which certainly meets these two criteria is known as valence shell electron pair repulsion theory (VSEPR) or sometimes by the names of its two most recent proponents and developers, Gillespie–Nyholm rules.

This is not the place to discuss VSEPR in detail; the rules and many examples can be found in any textbook of basic inorganic chemistry and, in most detail, in Gillespie and Hargittai (1991); see also Winter (1993). Nevertheless, there are some important trends in molecular structure which we shall want to address and for that purpose, it will be useful to present the basic rules of VSEPR and to note the reasons behind the common shapes which molecules adopt.

VSEPR as the name implies, is a set of rules with which to predict or rationalize the structures of molecules based on the repulsion between pairs of valence electrons; core electrons are completely ignored since their distribution around the nucleus is spherical and highly unpolarizable. We should bear in mind that the concept of 2-centre-2-electron bonds is used explicitly in this regard and hybridisation schemes can subsequently be employed once a structure is established, but we should recognise that VSEPR is a model for predicting the arrangement of atoms, not for describing details of electronic structure.

The geometries adopted for two to six pairs of electrons are linear, trigonal-planar, tetrahedral, trigonal-bipyramidal and octahedral respectively: illustrative examples where all of the electron pairs are bonding pairs are shown in Fig. 7.1.

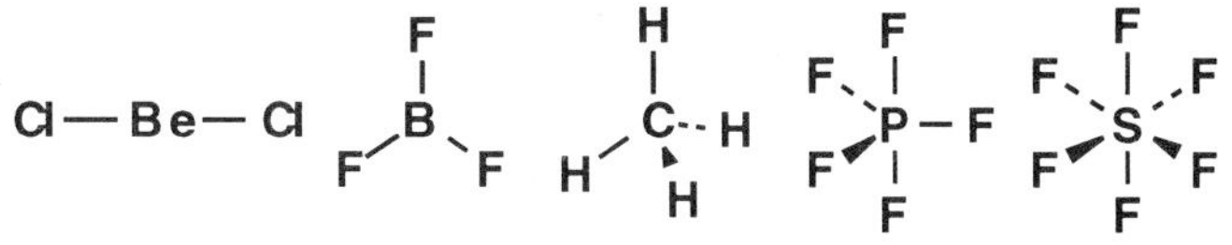

VSEPR is only really appropriate for s- and p-block molecules since for compounds of the d-block elements, the assumption of unpolarizable, spherical cores is not justified. For example, the molecule $[PtCl_4]^{2-}$ has 16 valence electrons or eight pairs, and four pairs can be assumed to be used in four Pt–Cl bonds. However, the molecule is square-planar rather than tetrahedral, as would be predicted for a p-block compound, since the remaining four non-bonding electron pairs do not form a spherical core or behave as normal, localized lone pairs. Where a spherical core is appropriate, as is the case for d^0, d^5 and d^{10} electron counts, VSEPR can be used.

More recent discussions of VSEPR focus more on electron pair domains (bond pair, lone pair etc) and the space they occupy rather than on electron pair repulsions. See, for example, Gillespie *et al.* *J.Chem.Educ.*, (1996), **73**, 622 and *Angew.Chem.*, (1996), **35**, 495.

Fig. 7.1 Examples of molecules with two to six bonding electron pairs.

Non-bonding electron pairs also occupy a distinct region of space and influence the structure accordingly. Thus NH_3 is pyramidal with three bonding pairs and one lone pair and H_2O is bent or V-shaped with two bond pairs and two lone pairs (Fig. 7.2). We should also note the general rule that non-bonding pairs occupy more space around a central atom than a bonding pair since they are associated only with that atom and are not 'stretched out' between two atoms as is the case with a bonding pair. Thus whereas in CH_4, the angles subtended at the central atom are 109.5°, i.e. those of the regular tetrahedron, the H–N–H angles in NH_3 are 107.5° and the H–O–H angle in H_2O is 104.5°. In terms of repulsion for electron pairs:

lone pair–lone pair > lone pair–bond pair > bond pair–bond pair

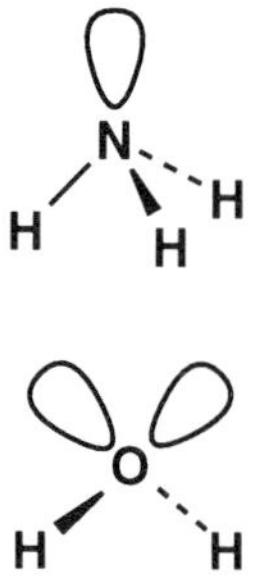

For a trigonal bipyramidal geometry, lone pairs always go into the equatorial sites as this minimizes the 90° contacts (each axial site has three 90° contacts to the equatorial positions whereas each equatorial site has only two 90° contacts to the axial positions). This is illustrated with the structures of SF_4 (equatorially vacant trigonal bipyramidal or disphenoidal), ClF_3 (T-shaped) and XeF_2 (linear) (Fig. 7.3).

For an octahedral geometry derived from six electron pairs, a molecule with five bonding pairs and one lone pair adopts a square-based pyramidal geometry, whereas for four bond pairs and two lone pairs, the latter occupy *trans* sites and a molecule like XeF_4 is square planar. These rules and the shapes are summarized in Table 7.1 where EX_nZ_m represents a central atom, E, surrounded by nX groups and m lone pairs, Z.

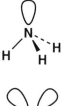

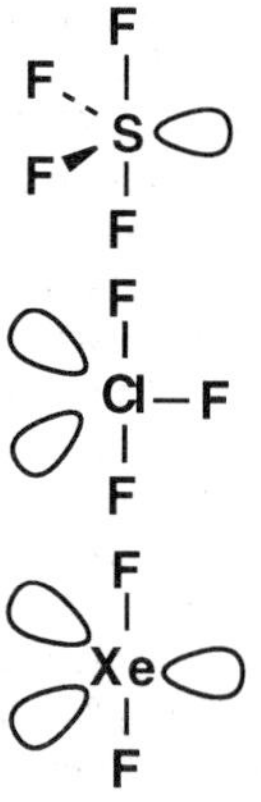

Fig. 7.2 The structures of NH_3 and H_2O showing the lone pairs.

Fig. 7.3 The structures of SF_4, ClF_3 and XeF_2 showing the lone pairs.

Table 7.1 Geometrical arrangements for EX_nZ_m molecules

No. of electron pairs	Electron pair arrangement	n	m	Type	Molecular shape	Examples
2	Linear	2	0	EX_2	Linear	$BeCl_2$
3	Trigonal planar	3	0	EX_3	Trigonal Planar	BF_3
		2	1	EX_2Z	V-shaped	$SnCl_2$
4	Tetrahedral	4	0	EX_4	Tetrahedral	CH_4
		3	1	EX_3Z	Trigonal Pyramidal	NH_3
		2	2	EX_2Z_2	V-shaped	H_2O
5	Trigonal bipyramidal	5	0	EX_5	Trigonal Bipyramidal	PF_5
		4	1	EX_4Z	Disphenoidal	SF_4
		3	2	EX_3Z_2	T-shaped	ClF_3
		2	3	EX_2Z_3	Linear	XeF_2
6	Octahedral	6	0	EX_6	Octahedral	SF_6
		5	1	EX_5Z	Square pyramidal	IF_5
		4	2	EX_4Z_2	Square planar	XeF_4

Molecules with double or triple bonds require an additonal rule which states that a multiple bond has the same gross stereochemical influence as a single bond. For example, in $POCl_3$ there are five bonding pairs, three P–Cl pairs and two pairs associated with the P=O double bond. Since the latter two pairs act as one, the geometry of $POCl_3$ is determined by four pairs and the molecule is tetrahedral. In general, multiple bond pairs exert a greater repulsion at the central atom than lone pairs or single bond pairs, and the interbond angles deviate from idealized values accordingly. A list of general structural types involving double bonds is presented in Table 7.2.

Table 7.2. Structural types for molecules involving double bonds

Total no. of electron pairs	No. of double bonds	No. of lone pairs	Shape	Example
3	1	0	V-shaped	NOCl
4	1	0	Trigonal planar	$COCl_2$
	2	0	Linear	CO_2
5	2	0	Trigonal planar	FNO_2
	2	1	V-shaped	SO_2
	1	0	Tetrahedral	$POCl_3$
	1	1	Trigonal pyramidal	$SOCl_2$
	1	2	V-shaped	FClO
6	3	0	Trigonal planar	SO_3
	2	0	Tetrahedral	SO_2Cl_2
	2	1	Trigonal pyramidal	$FClO_2$
	2	2	V-shaped	XeO_2
	1	0	Trigonal bipyramidal	SOF_4
	1	1	Disphenoidal	IOF_3
	1	2	T-shaped	$XeOF_2$

One final aspect of VSEPR which we will look at can be illustrated with the structure of PF_2Cl_3. This molecule is a simple derivative of PF_5 and has the expected trigonal bipyramidal structure, but the question remains as to how the atoms are arranged since a number of isomers is possible depending on which atoms are in the equatorial sites and which are axial. The structure observed is the one with both fluorine atoms in axial sites, and this observation turns out be quite general in that electronegative atoms or substituents prefer axial positions in a trigonal bipyramid. The origin of this site preference can be traced to the smaller effective size of electronegative atoms at the central element due to the resultant polarization of the bonding electron density away from the central atom. This enables electronegative atoms to more readily tolerate the axial site with its three 90° contacts (the opposite of why a lone pair prefers an equatorial site).

An alternative explanation is embodied in Bent's rule which concentrates more on the electronic origins of this site preference. Recall that in Section

A further application of Bent's rule is in rationalizing the precise structure of a compound like CF_2H_2 and related species. CF_2H_2 is tetrahedral, but the F–C–F angle is found to be less than the ideal tetrahedral angle whereas the H–C–H angle is greater. Normally, all four tetrahedral hybrids are considered as sp^3 but there is no requirement that they must be identical; some may have more s character and some less. Specifically, we may have two hybrids of the form sp^{3-x} and two of the form sp^{3+x}, the only condition being that the sum of all four hybrids must equal one s plus three p. In CF_2H_2 the fluorines bond through hybrids with a greater degree of p character (the less electronegative ones) and therefore have an interbond angle of less than the tetrahedral angle: in the limit where bonding was through pure p orbitals, the expected interbond angle would be 90°. In effect, of course, this distortion of the structure and the non-equivalence of the hybrids is driven by the fact that F and H have different electronegativities.

6.3.2, the hybrid orbitals appropriate for a trigonal bipyramid were discussed in terms of a dsp^3 set which could be broken down into equatorial sp^2 and axial dp subsets. These subsets are not equivalent and electrons will have different energies depending on which subset they are in. Specifically, electrons will be more tightly bound in the sp^2 subset rather than the dp set, since orbital energies are such that s is lower than p which is lower than d, as we saw in Chapter 1. Alternatively, we can assign an effective electronegativity to an orbital or hybrid, and it is clear that in the present example, the sp^2 set can be considered as more electronegative than the dp set since the electrons are more tightly held. The result is that an electronegative atom such as fluorine tends to bond to the least electronegative hybrid (given a choice), which in this case is dp, since there is less 'competition' for electrons thereby accounting for the observed axial site preference.

With the above overview of molecular structure according to VSEPR, it will now be useful to consider some of the periodic structural trends which are observed and to consider the various models which are used to account for them.

In general, hybrids with more s character are more electronegative; e.g. sp > sp^2 > sp^3.

7.2 Trivalent AX$_3$E compounds of group 15

Let us start by considering the trihydrides of the group 15 elements, EH$_3$. VSEPR predicts a trigonal pyramidal structure for these molecules since there is one lone pair as well as three E–H bond pairs. Moreover, as we saw above, the interbond angles in NH$_3$ are less than the ideal tetrahedral angle due to lone pair repulsions being stronger than bond pair repulsions. If we look at the trihydrides of the first four members of group 15 (BiH$_3$ is not stable), it is found that there is a steady decrease in the H–E–H angle as the group is descended as shown in Table 7.3.

In fact, decreasing interbond angles on descending a group is a general phenomenon in p-block chemistry where lone pairs are present; it is also seen in the group 16 dihydrides and in other examples discussed below. The explanation within VSEPR theory is straightforward since it is assumed that as the central atom, E, gets larger, the E–H bond pairs can be forced closer together by the lone pair before significant bond pair–bond pair repulsions arise. Moreover, the decreasing electronegativity of the element centre, as the group is descended, results in a relatively greater polarization of the bonding electron density towards the hydrogen atom. Polarization away from the central atom results in more space around that atom which also allows the E–H bond pairs to approach more closely. In actual fact, the group 15 elements are all more electronegative than hydrogen so the bonding electron density will tend to be polarized towards these atoms. Nevertheless, the extent to which this is so will decrease down the group.

Other important factors are apparent when we consider the trifluorides of group 15, interbond angles for which are shown in Table 7.4. The first point to note is that the F–E–F angles also decrease as the group is descended just as found for the hydrides and for much the same reasons. Note also that the interbond angle in SbF$_3$ is actually less than 90°, a point we will return to later, although some care must be exercised in these comparisons since SbF$_3$

Table 7.3 H–E–H bond angles (deg) for the group 15 trihydrides

Compound	H–E–H angle (deg)
NH$_3$	107.5
PH$_3$	93.2
AsH$_3$	92.1
SbH$_3$	91.6

Table 7.4 F–E–F bond angles (deg) for the group 15 trifluorides

Compound	F–E–F angle (deg)
NF$_3$	102.3
PF$_3$	97.8
AsF$_3$	95.8
SbF$_3$	87.3

has significant intermolecular Sb····F interactions in the solid state unlike the lighter congeners. A second point is that the interbond angle in NF_3 is less than that in NH_3. This may, at first, seem surprising since the fluorine atom is bigger than hydrogen but we must also consider that the N–F bond length is greater than the N–H distance and also that fluorine is much more electronegative than hydrogen. This latter aspect is important as the electron density will be substantially polarized towards the fluorine atoms and away from the nitrogen which will allow the N–F bond pairs to approach each other more closely at the nitrogen centre before inter-electron pair repulsions become significant.

The first problem we encounter is in considering PF_3 since, if the above arguments for NF_3 and NH_3 are correct, we might have expected the bond angles in PF_3 to be less than those in PH_3. Clearly this is not the case here or with the arsenic analogues. A similar situation is found with H_2O/F_2O and H_2S/F_2S and there is no easy way around this difficulty based on the arguments we have employed so far. Thus, whilst VSEPR is useful for predicting the gross structural type and for rationalizing trends within a homologous series, we must exercise some care when comparing angles in structures with different substituents especially where hydrogen is concerned.

Let us now further consider the trend to decreasing bond angles as group 15 is descended. We have seen illustrations of this trend with the element hydrides and fluorides but it is quite general for a range of substituents, and a similar effect is observed in group 16 for compounds of the general formula EX_2 and also with respect to the inter-equatorial bond angle in EX_4Z compounds.

We have already seen that VSEPR provides us with an initial level of explanation but we need also to consider the nature of the bonding. It is usually stated in textbooks of inorganic chemistry that the trend towards angles of 90° reflects a tendency for the heavier elements to bond only through p orbitals. In other words, that sp hybridization for the heavier elements does not readily occur. This explanation, as it stands, is not easily reconciled with the fact that the energy separation between the valence s and p orbitals decreases as the groups are descended, but it is consistent with the high s–p promotion energy and particularly with inefficient sp hybridization resulting from the different sizes of the s and p orbitals for the heavier elements (see Chapter 4).

It would therefore seem that bonding through pure p orbitals is an appropriate description for the heavier elements and that in, for example, triphenyl bismuth, $BiPh_3$, with C–Bi–C angles of 94°, the bismuth–carbon bonds are formed by overlap involving bismuth p orbitals and that the bismuth lone pair resides in a pure s orbital. This latter point is also consistent with the observed decrease in basicity of group 15 EX_3 compounds as the group is descended. Thus whilst PPh_3 readily forms complexes by acting as a Lewis base, the analogous chemistry for $BiPh_3$ is much less extensive due to s orbitals with poor directional properties not forming strong bonds which, coupled with the radial contraction of the heavier element s orbitals, leads generally to weak bonds and reduced basicity.

The resolution to the problem of the bond angles in the series of compounds NH_3, NF_3, PH_3 and PF_3 is probably associated, at least in part, with changes in the direction of bond polarity. Thus, if we consider electronegativities, the N–H bonds in NH_3 are polarized $\delta^- $ N–H δ^+ whereas for the other three compounds the polarization is such as to place the δ^+ charge on the nitrogen centre. This would lead to the N–H bond pairs taking up more space near the nitrogen centre and thus have the effect of widening the angles in NH_3 in relation to NF_3.

An interesting example of how this trend to smaller bond angles on descending a group affects compound stability is found with compounds of group 14 of the general formula R_8E_8 which have the cubane structure. Thus as the group is descended, the trend towards interbond angles of 90° results in the ring strain energy becoming progressively less.

The above argument is certainly one way of looking at things and is fairly satisfactory at explaining many features, although it does not readily account for why interbond angles of less than 90° should be observed as found, for example, in SbF_3. An alternative approach to this problem is to consider the nature of the molecular orbitals and how these can be used to rationalize the shapes of molecules.

Let us start by asking the question of why EX_3Z molecules are pyramidal from a molecular orbital point of view. In so doing we can start by considering a more symmetrical trigonal planar reference structure and constructing the molecular orbitals for this geometry. Fig. 7.4 shows a qualitative molecular orbital diagram for planar EH_3 and how the orbitals change in energy as the geometry is changed to trigonal pyramidal. This is known as a Walsh diagram; details of how it is constructed and the symmetry arguments involved are beyond the scope of this book but can be found in the books on molecular orbital theory listed in the bibliography.

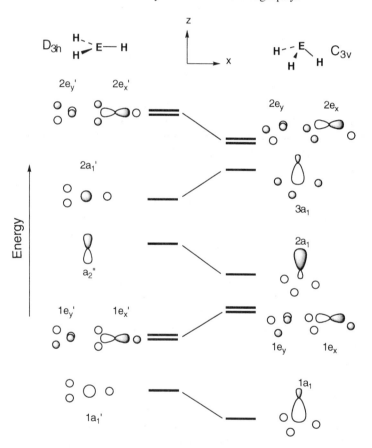

Fig. 7.4 An orbital interaction diagram for EH_3.

The important points to note are as follows. For a planar, 8-electron species, i.e. EX_3Z, the HOMO is the non-bonding a_2'' orbital in D_{3h} symmetry and the LUMO is the $2a_1'$ σ antibonding orbital. As the symmetry is lowered from planar to trigonal pyramidal, i.e. D_{3h} to C_{3v}, these two orbitals take on the same symmetry, a_1 in C_{3v}, and are therefore able to mix which results in a lowering of the energy of the HOMO (now $2a_1$) since it becomes slightly bonding, and a raising in energy of the LUMO (now $3a_1$) as it becomes more antibonding. This type of distortion is known as a second-order Jahn–Teller distortion and it is the lowering in energy of the HOMO which drives the distortion and results in 8-electron EX_3Z species having a trigonal pyramidal geometry rather than being trigonal planar. Note that in a 6-electron species such as BF_3, the doubly degenerate HOMO is lower in energy for the planar structure consistent with the observed geometry for this molecule.

We can take this argument a little further by considering the extent to which a distortion from planar to pyramidal will occur. In other words, what factors determine how pyramidal the molecule will become? In fact, this is governed largely by the energy separation between the HOMO and LUMO in the planar reference structure (the HOMO–LUMO gap). In situations where the energy separation is small, the orbitals can mix more effectively and the corresponding distortion to pyramidal will be large, whereas for a large energy gap, only small distortions will be observed.

If we now consider the energy levels for the group 15 EX_3Z species, the energy gap will decrease as the group is descended for two reasons. Firstly, the a_2'' HOMO will rise in energy as the element centre becomes less electronegative and secondly, the $2a_1'$ LUMO will fall in energy as the central atom becomes larger since the orbitals are getting bigger and more diffuse such that overlap is less efficient and the orbital is not so antibonding.

Not only does this model rationalize why pyramidality increases down the group but it also accounts for a related observation that pyramidal inversion energy barriers increase as the group is descended. In NH_3, the barrier to inversion at the nitrogen centre is about 25 kJ mol^{-1} whereas for PH_3 it is much larger at around 140 kJ mol^{-1}. This is directly related to the increasing pyramidality since, as a molecule becomes more pyramidal, the energy difference between $2a_1$ in C_{3v} and $2a_1'$ in D_{3h} becomes larger. In fact, this energy difference can be taken as a measure of the activation energy for inversion through a planar transition state.

Note that in all of these arguments, s–p mixing, if not explicit hybridization, is an important factor and also that there is nothing particularly special about an interbond angle of 90°. There is no reason why the distortion should not continue to angles less than 90° if this results in a further lowering of the energy of the HOMO.

Another trend which can be rationalised along similar lines concerns the increasing pyramidality found for the higher homologues of alkenes, $R_2E=ER_2$ (R = C to Sn). Thus whilst alkenes are generally planar, Fig. 7.5(a), the so-called disilenes, digermenes and distannenes increasingly adopt a *trans* bent geometry where each E centre is trigonal pyramidal rather than trigonal planar, Fig. 7.5(b). A molecular orbital scheme and associated second order Jahn-Teller effects can be employed to account for this trend, but also relevant are the energetics of the hypothetical fragments R_2E. Thus if

Another point to note is that for a given element, E, the molecule will become more pyramidal as the electronegativity of X increases since this will lead to a greater separation in energy between the orbitals of E and X which will lower the energy of $2a_1'$.

We should note that whilst pyramidal inversion energy barriers through a trigonal planar geometry do increase down the group in group 15, for the heavier congeners, an alternative process involving a T-shaped transition state has recently been found to be lower in energy.

Space does not permit a fuller discussion of the molecular orbital theory emloyed in the section. For this and a more in depth treatment of second order Jahn-Teller effects, the reader is refered to Albright, Burdett and Whangbo (1985) and Gimarc (1979).

the so-called singlet-triplet energy difference is small, as it is for carbon, two triplet fragments can combine to give a planar structure, as shown in Fig. 7.5(c), whereas for the larger singlet-triplet energy differences which occur as the group is descended (due to increasing s-p promotion energies, see Fig. 2.8), two singlet fragments combine to give the *trans* bent structure, Fig. 7.5(d).

The molecular orbital arguments presented above are quite general and can be used to rationalize most, if not all, of the structural trends observed in p-block chemistry. As is so often the case in chemistry, two apparently different theories or approaches are complementary rather than contradictory. The molecular orbital method is useful for understanding electronic structure but VSEPR is preferable in view of its simplicity, for quickly predicting structure.

7.3 The Zintl principle

As a final section on structure, we will consider a principle, which has been called the Zintl principle, that is helpful in identifying structural similarities between seemingly disparate compounds. Another name for this idea is the isoelectronic principle which is much more descriptive since it indicates that the key to structural similarity is to be found in the number of electrons (specifically valence electrons). Thus, if a set of compounds have the same overall number of valence electrons they are likely to have similar structures provided that the elements do not differ widely in electronegativity or polarizability.

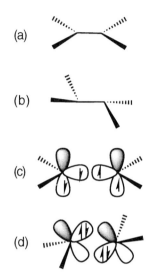

Fig. 7.5 Planar (a) and *trans* bent (b) forms of $R_2E=ER_2$ and representations of the overlap of two triplet (c) and two singlet (d) R_2E fragments. A triplet state is one where there are two unpaired electrons whereas these electrons are paired in a singlet state.

13–15 clearly refers to the atomic groups; this is equivalent to the Roman numeral designation III–V derived from the old group numbering scheme (i.e. main groups 3 and 5).

As a first example, we can consider polymeric compounds which have the cubic diamond structure. These include the so-called III–V (or 13–15) compounds boron nitride (BN), aluminium phosphide (AlP) and gallium arsenide (GaAs), the II–VI (or 12–16) compounds zinc selenide (ZnSe) and cadmiun telluride (CdTe) and the I–VII (or 11–17) compounds copper(I) bromide (CuBr) and silver iodide (AgI) all of which have four bond pairs per atom and no lone pairs.

For more ionic systems, an example is the structure of magnesium diboride, MgB_2. In this material the magnesium is present as Mg^{2+}, the two electrons lost having been transferred to the boron atoms. Since there are two borons per magnesium, each boron is present as B^- which is isoelectronic with carbon and the structure comprises planar, hexagonal sheets of boron atoms isostructural with the sheets in graphite (Fig. 3.3). Similarly, in lithium arsenide, LiAs, the arsenic is present as As^- and is isostructural with grey selenium (Fig. 3.6(c)) whilst in calcium disilicide, $CaSi_2$, the silicons are present as Si^- and are isostructural with orthorhombic black phosphorus (Fig. 3.5(c)). Note that in calcium carbide, CaC_2, the C_2 unit is present as a discrete C_2^{2-} ion isostructural with N_2 which illustrates the prevalence of π-bonding amongst the 2p elements. As a final example, we may note that the thallium atoms in the structure of NaTl are present as Tl^- and are arranged in a diamond-like, three-dimensional network. Many ternary compounds can also be rationalized according to these ideas.

Bibliography

General inorganic chemistry texts

1. Shriver, D. F., Atkins, P. W., and Langford, C. H. , (1994). *Inorganic chemistry*, (2nd edn). Oxford University Press.
2. Huheey, J. E., Keiter, E. A., Keiter, R. L. (1993) *Inorganic chemistry*, (4th edn). Harper and Row, New York.
3. Cotton, F. A., Wilkinson, G. and Gaus, P. (1987). *Basic inorganic chemistry*, (2nd edn). Wiley, New York.
4. Cotton, F. A. and Wilkinson, G. (1988). *Advanced inorganic chemistry,* (5th edn). Wiley, New York.
5. Jolly, W. L. (1991). *Modern inorganic chemistry*, McGraw–Hill, New York.
6. Greenwood, N. N. and Earnshaw, A. (1984). *Chemistry of the elements*, Pergamon Press, Oxford.

Texts dealing with structural inorganic chemistry

1. Wells, A. F. (1984). *Structural inorganic chemistry*, (5th edn). Oxford University Press.
2. Gillespie, R. J. and Hargittai, I. (1991). *The VSEPR model of molecular geometry*, Allyn and Bacon, Boston.
3. Müller, U. (1993). *Inorganic structural chemistry*, Wiley, New York.
4. Alcock, N. W. (1990). *Bonding and structure*, Ellis Horwood, Chichester.

Texts dealing specifically with main group chemistry

1. Massey, A. G. (1990). *Main group chemistry*, Ellis Horwood, Chichester.

Texts dealing with the understanding of Inorganic Chemistry principles and molecular orbital theory

1. Owen, S. M. and Brooker, A. T. (1991). *A guide to modern inorganic chemistry*, Longman, Harlow.
2. Barrett, J. (1991). *Understanding inorganic chemistry*, Ellis Horwood, Chichester.
3. Dasent, W. E. (1965). *Nonexistent compounds*, Edward Arnold, New York.
4. Winter, M. J. (1993). *Chemical bonding*, Oxford University Press.
5. Albright, T. A., Burdett, J. K., Whangbo, M. H. (1985). *Orbital interactions in chemistry*, Wiley, New York.
6. Gimarc, B. M. (1979). *Molecular structure and bonding*, Academic Press, New York.
7. Mingos, D. M. P. (1997). *Essential trends in inorganic chemistry*, Oxford University Press.

Books containing useful data

1. Emsley, J. (1989). *The elements*, Oxford University Press.

Index